INTRODUCTION TO GENERAL RELATIVITY AND THE COSMOLOGICAL CONSTANT PROBLEM

INTRODUCTION TO GENERAL RELATIVITY AND THE COSMOLOGICAL CONSTANT PROBLEM

MARCELO SAMUEL BERMAN

INSTITUTO ALBERT EINSTEIN (CURITIBA)

Nova Science Publishers, Inc.

New York

For permission to use material from this book please contact us:
Telephone 631-231-7269; Fax 631-231-8175
Web Site: http://www.novapublishers.com

NOTICE TO THE READER

Library of Congress Cataloging-in-Publication Data Berman, Marcelo Samuel.
Introduction to general relativity and the cosmological constant problem / Marcelo Samuel Berman.
p. cm.
Includes index.
ISBN 13 978-1-59454-717-1
ISBN 10 1-59454-717-3
1. General relativity (Physics) 2. Cosmology. I. Title.
QC173.6.B47 2004
530.11–dc22 2006000430

Published by Nova Science Publishers, Inc. ✤ New York

For Albert, Clarissa, Geni, Gloria, Miguelina, Paula and Rosa.
In honour of Dr. Paulo de Tarso Monte-Serrat
and Prof. Clion Dória.

"There are things that happen whether you know all about them or not.

A person who does not know anything about the law of gravity suffers from the same effects as another one who knows all about it.

Gravity is not at all concerned with whether you know about its existence.

If a physicist and an ignorant person both jump down into an abyss, the result will be the same: a human paste".

Lair Ribeiro, in *Feet on the Ground, Head in the Stars*. (2002)

Contents

Preface and Introduction

This textbook is a simple introduction to General Relativity Theory (GRT), and the cosmological constant (lambda – Λ) problem designed for senior undergraduates, or first year graduate students in Theoretical Physics. It is the first elementary account of GRT, and Cosmology, to address the lambda problem. Should one prefer scientific theories that contain beautiful equations, rather than ugly ones? I am against prejudices, but beautiful equations are fundamental!!!

Just imagine the reaction against a guru-cosmologist, in the early 90's of last century, telling his colleagues that first, Einstein-deSitter's model is no good; second, that cosmologists should have defined an acceleration parameter, rather than a deceleration one; third, that the Universe has been lambda–dominated, for the most part of its life, and still is so; fourth, that Einstein's greatest "blunder", could be, discarding lambda, but not, introducing it; fifth, that in fifteen years "from now", a publisher would sign a contract for two textbooks, designed for beginners, one, on the lambda term, the second, dealing quite extensively with scalar tensor cosmologies, and that in those books there were cosmological models which considered the fundamental constants, like lambda, fine structure "constant", and the speed of light, to be time-varying. Sixth, that the author of both books would suggest that the Universe resembled a huge white-hole, and this author would be not alone (a second one, Paul Wesson, would publish at the same time a book on pentadimensional physics, with a second different proof on that); seventh, that Berman would claim that the neutrinos' mass was time-varying, and Physical Review Letters would publish a similar claim, by Kaplan (2004).

I do not know how to describe the sort of statements that will be common in fifteen years from now, but certainly if I were a guru, like the one above, I would face discredit if I talked about the year 2020's physics.

I was 24 years old, when Flavio S. de Lacerda, a Brazilian Minister of Education declared that I was "a great scientist"; but at the same time, I was facing repression supported by the military regime who held power in Brazil. The result was that I had my career brutally interrupted for more than a decade. Were not for this fact, the present book would have probably been written with more maturity.

This is not an exhaustive treatise, and the main goal is to supply information that might help the reader in following the solutions for the 10^{120} or so orders of magnitude that differentiate the value of the cosmological constant, in the very early Universe, and its value today. Recent developments include the realization that a tiny but non zero cosmological constant is responsible for the accelerating present Universe. As far as I know, the only reasonable mechanism to attain the desired tiny value for the cosmological constant today, while keeping a huge value in the very early Universe is to accept that we are in face of a cosmological variable term, that, for instance, decreases, with the square of the age of the Universe or, else, with the square of the radius of the Universe. Such solution had been made by several authors, including Bertolami and Berman, while dealing with a Brans-Dicke model of the Universe endowed with a cosmological term, in the late 80's. Soon afterwards Chen and Wu suggested the R^{-2} dependence, for Λ, followed by some papers of mine. I believe that the lambda-term should be the element of Physics that may enable us to make a bridge between GRT and Quantum Mechanics:both theories deal with the energy of the vacuum ("zero-point " energy).

The importance of the cosmological constant today lies basically on another assumption of Cosmological Theory: if the inflationary model of the Universe is correct, we expect the density of the Universe to be critical. This led cosmologists to assume that the energy of the vacuum responds for a large amount of the energy of the Universe. As was stressed by the Russian cosmologist Y. Zel'dovich in the year 1974, the non null energy of the vacuum implies a non null cosmological constant. The acceleration of the Universe, and the variation of the fine-structure "constant" are here given some space. The news accounting for the 2006 experimental evidence that the Universe has been Λ – driven for a long period in its history, is not a surprise to me; this fact was evident from my treatments on the Machian Universe, where it was derived, from the zero-total energy of the Universe conjecture, that each particular form of energy density, would keep its relative importance towards the total energy density, during all the Universe's life. (Berman, 2006; 2006 b).

The first two chapters of this book, deal with the mathematical foundation of GRT; just the bare rudiments of Riemannian geometry and tensor calculus that are needed for understanding this book are touched upon. Next we deal with GRT leading to Einstein's field equations and then, the Schwarzschild solution that represents a black hole or the solar system. Further chapters introduce the minimum possible account of Relativistic Cosmology in order to deal with our goal which is the solution of the cosmological constant problem. A more complete treatment of Relativistic Cosmology will be found in the companion book by Berman (2007). Several cosmological models are then studied along with the imposition of a variable cosmological constant and also with a time-varying fine-structure "constant". Sections 10.1 and 10.2 require more mathematical sophistication, and could be skipped in a first reading.

The new century has brought, in its five first years, two important discoveries: the first, is the confirmation of the Lense-Thirring metric effects, reported by Ignazio Ciufolini and collaborators (2004), and the first general relativistic treatment of the rotational speeds in galaxies, by Fred Cooperstock and S. Tieu (2005), that may well bury some concepts over the dark matter. The Lense-Thirring metric and other developments, as well as Cooperstock's approach to dark matter and rotational speeds in galaxies, are dealt in the Appendices.

The References and Suggested Bibliography is constituted of three Parts: Parts RI and RII, refer to papers and books by John D. Barrow, which, as it seems to me, make for the most important set of publications from one single author, in General Relativity and Cosmology. Part RIII lists other authors' contributions.

My thanks are due to Professor Fernando de Mello Gomide, who coauthored two books with me (Berman and Gomide, 1987; Gomide and Berman, 1988), parts of which are included in the present one. I thank Professor Murari Mohan Som for reading the preliminary version of this book and for advices given by him to me during the last twenty years. Both Prof. Gomide and Prof. Som were my advisors and intellectual mentors and now,are friends and colleagues; Prof. Adonai Sant'Anna checked all the equations.

My colleagues and friends, Professors Antonio F.F. Teixeira, Mauro D. Tonasse and Nelson Suga, helped me with the gathering of useful paper reprints. All the text deals only with Classical Theory (i.e., non-Quantum); it reflects some of the work done by many people, including myself, in Relativistic Cosmology. The only exception lies in a brief use of Heisemberg's uncertainty principle in order to "derive" black holes temperature formula.

I take the opportunity to thank my colleagues Herman Cuesta, Homero Santiago Maciel, Mauro Nagashima, Odylio D. de Aguiar, Ramiro Wahrhaftig, Rubens de Melo Marinho Jr., and Tobias Federico. Newton da Costa, Full Professor of the University of São Paulo, one of the most outstanding mathematical logicians in Latin America, my former high school teacher of mathematics, gave me a lot of his most friendly incentive.

I remain grateful to Professors Fred Cooperstock, James Ipser and S. Gottesman, and also to the friendly colleagues I had during the several months I stayed at the University of Alabama. Maurice Bazin, one of the two authors that are alive, out of the three (R.J. Adler, M. Bazin and M. Schiffer) who wrote the unsurpassed introductory book on General Relativity, was gentle enough in many occasions. My public thanks go also to Lucasian Professor Stephen Hawking and his staff, who allowed me to prepare and edit the Brazilian version of "The Illustrated a Brief History of Time". Prof. H. Fleming exerted fruitfull influence on the author, too.

Most of the usual cosmological models employed by cosmologists, are of the constant deceleration parameter type. I have the credit for having derived them (Berman, 1983; Berman and Gomide, 1988), at a time when no one attached a serious reason for their existence. This textbook deals with mathematical cosmological models, not with experimental Cosmology. Being this an openly didactic book, we have inserted several repetitive passages which, we hope, will not bore the more competent readers. Repetition is necessary if you want to instruct beginning students. Of course, you would not be so redundant in a doctoral thesis. We have sacrificed mathematical rigor, in order that readers with different backgrounds, would not become exaggeratedly frustrated.

We change freely from Latin to Greek indices and vice-versa, without penalty: the reader shall easily decide in each particular case if the index runs from 0 to 3 or from 1 to 3. At times, we suppose that the constants G and c have unit value; in other times, we include these constants into the text. Later, while reading scientific papers on the subject, some difficulties of this kind will, then, be easily overcome. The pre-requisites for this book are University Physics (at the level of Halliday et al., 2005; Ohanian, 1985), and Advanced Calculus (Wrede and Spiegel, 2002).

In parallel to this humble book, we prepared, a companion one, which is being published simultaneously, under the title "*Introduction to General Relativistic, and Scalar Tensor Cosmologies*". An additional textbook, "*A Primer in Black Holes, Mach's Principle, and Gravitational Energy*", will appear in a few months. During proof-reading (November, 2006), I had some chance of changing the contents of this preface and beyond.

I owe a special gratitude debt with Mr. Marcelo F. Guimarães who has collaborated with TEX, throughout this text and many others. Clarissa Berman has also revised the spelling.

We are not native English speakers, so I ask the reader's indulgence with the eventual and unavoidable language mistakes.

I take the sole responsibility for the errors that may be in the text. I ask the reader to blame me for anything he does not like; but he should praise Professors F.M. Gomide, and M.M. Som in the opposite case.

Last, but not least, the author extends his thanks, to Frank Columbus, President of Nova Science Publishers, who invited me to become a Nova author. His staff had an important role in all the phases of the production, and to them I give part of the credit for this book.

Prof. Marcelo Samuel Berman
Curitiba - November, 2005.
msberman@institutoalberteinstein.org

Part I

Mathematical Preliminaries

Chapter 1

Tensors in Amorphous Spaces

1.1. Introduction

Let $x^1, x^2, ..., x^N$ be N variables. Each set of N possible values for these variables is called a point. The variables are called coordinates, and the totality of points corresponding to all the possible values of the coordinates, defines an N-dimensional space. Other names usually given are "manifolds", or "hyperspaces", which means basically the same thing. The particular case $N = 4$ is usually well known for its role in the Theory of Relativity, where the four dimensional space is called spacetime.

The Tensor Calculus is a tool which establishes a mathematical language valid for any system of coordinates, and not only in a particular one. As it will become clear later, equations written with the help of tensors benefit from this property. Nobel prize winner Albert Einstein made recourse to this tool (Tensor Calculus) in order to establish relations valid for any observers in arbitrary relative motion, by stating the Principle of Covariance of the physical laws. It is very evident that a given physical law may not be written in tensor form and nevertheless be covariant. Carmeli [1982] synthesized three possible statements for the Principle of Covariance, which are not exactly equivalents:

1st) all coordinate systems are equally good for expressing the Physical Laws, and all of them should be treated equivalently;

2nd) the Physical Laws must be expressed in tensor form in a four dimensional Riemannian spacetime;

3rd) each Physical Law must possess the same form in all coordinate systems.

In this Chapter we introduce tensors and we demonstrate that each tensor equation, if valid in a particular coordinate system, must be valid in any other one.

1.2. Vectors and Tensors

Let S and S' represented two coordinate systems, arbitrarily chosen, whose coordinates are represented respectively by x^i and x'^i, where i varies from 1 to N, in N dimensional spaces.

If we compare both systems of coordinates we will find between S and S', relations of the type:

$$x'^i = F^i(x^j), \tag{1.2.1}$$

where | $i = 1, \ldots, N$

| $j = 1, \ldots, N$
| F^i = one to one function, continuous with continuous derivatives of any order.

Taking the differential of relation (1.2.1) we obtain by the rules of differential calculus:

$$dx'^i = \sum_{j=1}^{N} \frac{\partial x'^i}{\partial x^j} dx^j. \tag{1.2.2}$$

From now on, we shall adhere to Einstein's sum convention, under which when one index is repeated twice in the same term of an equation, a sum over all values of that index is assumed. For example, we shall re-write (1.2.2) as:

$$dx'^i = \frac{\partial x'^i}{\partial x^j} dx^j. \tag{1.2.3}$$

Repeated indices are called in English language as "dummies".

We now introduce the symbol δ^i_j, called Kronecker's delta. It is defined by the following relations:

$$\delta^i_j = 1 \quad \text{if} \quad i = j,$$

$$\delta^i_j = 0 \quad \text{if} \quad i \neq j.$$

A very useful mathematical relation is the following one:

$$\frac{\partial x^i}{\partial x^j} = \delta^i_j. \tag{1.2.4}$$

The above relation expresses that different coordinates are independent of one another. Applying a well known rule from Calculus, we can write from (1.2.4) that:

$$\frac{\partial x^i}{\partial x'^j} \frac{\partial x'^j}{\partial x^k} = \delta^i_k. \tag{1.2.5}$$

Contravariant Vectors

Relation (1.2.3) defines the transformation law for the components of the displacement vector (dx^i) between two infinitesimally nearby points. Each set of N quantities V^i which transform likewise, is called the set of components of a contravariant vector, so that the prototype vector V'^i in system S' is related to V^j in S by

$$V'^i = \frac{\partial x'^i}{\partial x^j} V^j. \tag{1.2.6}$$

Contravariant Tensors of 2^{ND}Rank

Let T^{ij} be a set of N^2 quantities which transform according to the law

$$T'^{ij} = \frac{\partial x'^i}{\partial x^k}\frac{\partial x'^j}{\partial x^l}T^{kl}. \tag{1.2.7}$$

The above set T^{ij} is said to represent a contravariant tensor of 2nd rank. By the same token, higher order tensors could be defined. A closed analysis by the reader should suffice to check that by multiplying two contravariant vectors U^i and V^j, we obtain a 2nd rank contravariant tensor, and so on. However not always one could decompose a given tensor of superior order into the product of vectors. The reader may check also that by the given definitions, a contravariant vector is a tensor of first rank. The rank zero tensor is called an invariant; let ϕ and ϕ' represent an invariant in S and S' coordinate systems: obviously,

$$\phi = \phi'. \tag{1.2.8}$$

Covariant Vectors

Suppose that $\phi = \phi\ (x^i)$, is an invariant in each point of space, but a function of the coordinates x^i. From Calculus comes the relation

$$\frac{\partial \phi}{\partial x'^i} = \frac{\partial \phi}{\partial x^j}\frac{\partial x^j}{\partial x'^i}. \tag{1.2.9}$$

The reader should not have any trouble for identifying the "vector" $\frac{\partial \phi}{\partial x^j}$ as the gradient of ϕ. Each set of quantities V_i that transforms according to the gradient, i.e., according to (1.2.9), so that:

$$V'_j = \frac{\partial x^i}{\partial x'^j}V_i, \tag{1.2.10}$$

is called the set of components of a "covariant" vector, also called covector. As is clear in our notation, we adopt the convention of representing superior indices as a sign of contravariancy, while down indices represent covariance.

Covariant Tensors of 2^{nd}Rank

A set of quantities T_{ij} is called the set of components of a 2nd rank covariant tensor if when transformed to the system S' we obtain the set T'_{ij} defined as:

$$T'_{ij} = \frac{\partial x^k}{\partial x'^i}\frac{\partial x^l}{\partial x'^j}T_{kl}. \tag{1.2.11}$$

Covariant tensors of higher order are defined analogously. The reader can check that the covariant tensor of 1st rank is a covector. The invariant is a covariant tensor of zero rank. In other words, the invariant is both covariant and contravariant. Like in the contravariant tensor case, the product of a covariant vector with components V_i with another covariant vector U_j yields a covariant tensor of 2nd rank but it is useful to remember that not all covariant tensors of 2nd rank can be decomposed into the product of two covariant vectors.

Mixed Tensors

Consider the product of a contravariant vector V^i with a covariant vector U_j:

$$T^i_j = V^i U_j. \quad (1.2.12)$$

In another coordinate system T^i_j would be transformed into:

$$T'^i_j = \frac{\partial x'^i}{\partial x^k}\frac{\partial x^l}{x'^j} T^k_l. \quad (1.2.13)$$

We say that the set T^i_j represents the components of a 2nd rank mixed tensor (one time covariant and other time contravariant).

In a similar way, mixed tensors of rank $(m+n)$ can be defined, m times covariant and n times contravariant.

We invite the reader to show that Kronecker's delta is a mixed tensor of 2nd order. Hint: Notice the identity $\delta^r_s V^s_j U^k \equiv V^r_j U^k$. The general rule is that when in a product with δ^r_s you can put $r = s$ and eliminate the delta.

Solved Problem

A covariant vector has components in the three-dimensional cartesian system of coordinates (x, y, z) given by:

$$\begin{aligned} T_x &= xyz \\ T_y &= y \\ T_z &= x. \end{aligned} \quad (1.2.14)$$

It is asked to obtain the spherical components (T_r, T_θ, T_ϕ).

Solution:

In the notation of tensor calculus,

$$\begin{aligned} x^1 &= x \quad x'^1 = r \\ x^2 &= y \quad \text{and} \quad x'^2 = \theta \\ x^3 &= z \quad x'^3 = \phi. \end{aligned} \quad (1.2.15)$$

According to the data given in the problem,

$$\begin{aligned} T_1 &= x^1 x^2 x^3 \\ T_2 &= x^2 \\ T_3 &= x^1. \end{aligned} \quad (1.2.16)$$

The relation which express the cartesian coordinates as function of the spherical ones can be written as:

$$x^1 = x'^1 sen x'^2 . \cos x'^3$$

$$x^2 = x'^1 sen x'^2 . sen x'^3 \qquad (1.2.17)$$
$$x^3 = x'^1 \cos x'^2 .$$

We now recall the definition of a covariant vector, namely:

$$T'_j = \frac{\partial x^k}{\partial x'^j} T_k .$$

$$\therefore$$

$$T_r \equiv T'_1 = \frac{\partial x^1}{\partial x'^1} T_1 + \frac{\partial x^2}{\partial x'^1} T_2 + \frac{\partial x^3}{\partial x'^1} T_3$$

$$T_\theta \equiv T'_2 = \frac{\partial x^1}{\partial x'^2} T_1 + \frac{\partial x^2}{\partial x'^2} T_2 + \frac{\partial x^3}{\partial x'^2} T_3$$
$$T_\phi \equiv T'_3 = \frac{\partial x^1}{\partial x'^3} T_1 + \frac{\partial x^2}{\partial x'^3} T_2 + \frac{\partial x^3}{\partial x'^3} T_3 .$$

On performing the calculations we obtain the solution of the problem. For instance:

$$\begin{aligned} T_r &= sen x'^2 \cos x'^3 \; x^1 x^2 x^3 + sen x'^2 sen x'^3 \; x^2 + \cos x'^2 \; x^3 = \\ &= \text{sen}\;\theta \cos\phi \; (r\; sen\theta \cos\phi)\; (r\; sen\theta\; sen\phi)\; (r \cos\theta) + \\ &+ \text{sen}\;\theta\; sen\phi\; r\; sen\theta\; sen\phi + \cos\theta\; (r \cos\theta) = \\ &= r^3 (sen\theta)^3 \; (\cos\phi)^2 \; sen\phi \cos\theta + r\; (sen\theta)^2\; (sen\phi)^2 + r\; (\cos\theta)^2 . \end{aligned}$$

1.3. Algebraic Operations with Tensors

Two tensors of the same rank and type may be added (or subtracted), yielding a new tensor of the same rank and type. The idea is that you only can sum bananas with bananas and apples with apples. By the type of the tensor we mean the precise number of contravariant indices and covariant indices. Example: given two 3rd rank tensors of the same type, say, $A^i_{\;jk}$ and $B^i_{\;jk}$ we obtain a new tensor

$$C^i_{\;jk} = A^i_{\;jk} + B^i_{\;jk} .$$

From the position of the indices it is clear that the sum is of the same type and rank. It has to be noticed once and for all that we speak of a tensor by referring to the set of its components which we represent by a typical member like $C^i_{\;jk}$ that stands for the set.

FUNDAMENTAL THEOREM OF TENSOR CALCULUS:

Given a tensor equation valid in one system of coordinates, when we transform to another system of coordinates, both sides of the equation may change but we still get a valid equation.

Demonstration:

Let the given equation be valid in system S:

$$A^i_j = B^i_j.$$

Transforming into the coordinate system S' it is obvious that both sides transform in the same way, verbi gratia, like a mixed tensor of 2nd rank (with other type of tensors the reasoning would not be any different), so that:

$$A'^i_j = B'^i_j.$$

Simply stated, a tensor equation is valid in any system of coordinates. The mathematical reason behind this theorem is that tensor transformations are linear and homogeneous. Immediately one can conclude that any equation of the type $C^i_j = 0$ implies that

$C'^i_j = 0$, provided that we are dealing with a tensor.

(Remember that not all matrices of 2nd rank transform like tensors). In the above example if you write

$C^i_j = A^i_j - B^i_j = 0$, it is clear that $A'^i_j = B'^i_j$.

Exterior Product

The exterior product of two tensors was introduced earlier in the particular case when we commented that the product $V^j U_i$ of two vectors makes a 2nd rank tensor. A new tensor is always obtained which has a rank given by the sum of the ranks of the involved tensors in the exterior product, by the multiplication of the components of two or more tensors. As no summing is implicit in exterior product, care must be taken in order to represent each multiplying factor with different literal indices without repetition. For instance, suppose that we are given two tensors, say, A^r_{st} and B^{ik}_j. We multiply exteriorly to obtain a new tensor: C^{rik}_{stj}.

The value of each component of the new tensor will be given by:

$$C^{rik}_{stj} = A^r_{st} B^{ik}_j.$$

Contraction

Let a given mixed tensor be contracted which means that we write one index of covariant type equal to a given contravariant index: we claim that the resultant tensor has a rank two units less than the original rank. In other words, the contracted tensor has a rank which results from computing the left untouched indices.

Demonstration: consider the tensor A^i_{jkl}. By definition the law of transformation is:

$$A'^i_{jkl} = \frac{\partial x'^i}{\partial x^r}\frac{\partial x^s}{\partial x'^j}\frac{\partial x^t}{\partial x'^k}\frac{\partial x^p}{\partial x'^l}A^r_{stp}.$$

On contracting, i.e., writing $i = j$ we are left with:

$$A'^{i}_{ikl} = \left(\frac{\partial x'^i}{\partial x^r}\frac{\partial x^s}{\partial x'^i}\right)\frac{\partial x^t}{\partial x'^k}\frac{\partial x^p}{\partial x^l}A^r_{stp} = \frac{\partial x^t}{\partial x'^k}\frac{\partial x^p}{\partial x^l}A^r_{rtp},$$ where we made use of (1.2.5).

Q.O.D.

Interior Product

We define the interior product of two tensors by the exterior product followed by a contraction.

Tests of Tensor Character

Given a certain set of quantities, we would like to test the set in order to check whether we are in face of a tensor. The direct test is to check how the given set of quantities behaves under a coordinate transformation, say, from S to S'. Nevertheless, sometimes it is more useful to apply the so called Quotient Theorem which allows an indirect test: "if the interior product of the set "X" with an arbitrary tensor is a tensor, then X is a tensor". We leave it to the reader to demonstrate this theorem for a particular tensor like a 2nd rank contravariant. In other cases, the proof runs equally well.

1.4. Final Observations

1) **Transitivity**: Besides linearity and homogeneity the transformation property of tensors is endowed with transitivity. By transitivity we mean the following: if in a transformation from the coordinate system S to S' a given set X transforms according to the tensor definition, and the same happens from coordinate system S' to a third system S'' we are sure that the tensor character will be assured again. If this does not happen, the tensor definition would not be transitive. Let us demonstrate it for a 2nd rank contravariant tensor T^{ij} : the hypotheses are:

$$T'^{ij} = \frac{\partial x'^i}{\partial x^k}\frac{\partial x'^j}{\partial x^l}T^{kl}. \tag{1.4.1}$$

and,

$$T''^{s} = \frac{\partial x''^r}{\partial x'^i}\frac{\partial x''^s}{\partial x'^j}T'^{ij}. \tag{1.4.2}$$

The thesis is:

$$T'' = \frac{\partial x''^r}{\partial x^k}\frac{\partial x''^s}{\partial x^l}T^{kl}. \tag{1.4.3}$$

We demonstrate easily that substituting (1.4.2) into (1.4.1), and by applying twice the relation (1.2.5) we obtain (1.4.3).

2) **Tensorial Field**: The tensors defined in this Chapter refer to points in an amorphous space (Synge and Schild (1969)), what means that in this space there is no definition of distance between neighborhood points (the metric of the space is not necessarily defined

yet). Everything which is valid for isolated points in space, may be valid for a continuum, in which case we define a tensor field.

3) **Symmetry and Anti-symmetry**: A tensor of any order is said symmetric relative to a pair of indices of the same type, if the numerical value is not altered by an interchanged of those indices.

Example: if $T_{pq}^{rsl} = T_{pq}^{srl}$ then the tensor is symmetric relative to the first two contravariant indices.

We define an anti-symmetric tensor, relative to a pair of indices of the same type, if upon exchange of those indices, the components change the algebraic sign.

Example: suppose $T_{pq}^{rsl} = -T_{qp}^{rsl}$; then the tensor is anti-symmetric relative to the two covariant indices.

Anti-symmetric tensors are also called skew-symmetric tensors.

The reader may easily demonstrate that symmetry or anti-symmetry properties are conserved in a transformation of coordinates, for a tensor. This does not imply that an equality of the type:

$$T_{pq}^{rsl} = T_{rq}^{psl},$$

will be kept under a transformation of coordinates. As an exercise the reader should show that the above is true.

In the next chapter use will be made of the following Theorem: "let $\Phi = g_{ij}X^iX^j$ be an invariant while X^i and X^j are arbitrary contravariant vectors and g_{ij} is a set of symmetric quantities in all system of coordinates. Then, g_{ij} stands for a 2nd rank covariant tensor".

4) **Riemannian Spaces:** For the reader versed in differential geometry we want to advance the idea that we shall need to particularize the otherwise amorphous space of this Chapter into a metric Riemannian space which is torsionless, where the affinities are the Christoffel symbols and there is a metric tensor of 2nd rank which is symmetric. The whole story is left for next Chapter.

References for Chapter 1

At this stage, the reader may benefit from consulting more advanced treatises like Weinberg (1972), Adler, Bazin and Schiffer (1975) or MTW (1973). These references are basic and highly recommended reading.

Chapter 2

Tensors in Riemann Spaces

2.1. Metric

In Euclidean Geometry, it is true that the distance between infinitesimally nearby points is an invariant. Choosing Cartesian orthogonal coordinates, we shall have:

$$ds^2 = dx^i dx^i \quad = \quad \text{invariant}, \tag{2.1.1}$$

where a summation from 1 to 3 is implied by the repeated index i.

In curvilinear coordinates, ds^2, called in general the metric form or the fundamental space form or the square of the line element, is given by an expression of the type:

$$ds^2 = g_{ij} dx^i dx^j. \tag{2.1.2}$$

For instance, in spherical coordinates, we shall have:

$$\begin{aligned} g_{11} &= 1, \\ g_{22} &= r^2, \\ g_{33} &= r^2 sen^2\theta, \\ g_{ij} &= 0, \text{ for } i \neq j. \end{aligned} \tag{2.1.3}$$

The 2nd rank covariant tensor, g_{ij}, is called covariant metric tensor or the fundamental space tensor. Notice that we may consider, a priori, the tensor g_{ij} to be symmetric, having in mind the commutativity of the product $dx^i dx^j$, and the subsequent verification that in expressions (2.1.2), only occur sums of the type

$$(g_{ij} + g_{ji}).$$

Notice also that the proof that g_{ij} is a 2nd rank covariant tensor is obtainable from last Chapter theorem from Section I.4 where we take

$$\Phi = ds^2,$$

$$X^i = dx^i.$$

We say that the metric is indefinite whenever it may assume positive, negative or null numerical values. In the case of General Relativity Theory, as well as in the particular case where there is no gravitational field (Special Relativity) this can indeed happen. Recall Minkowski's spacetime metric:

$$ds^2 = -\left[dx^2 + dy^2 + dz^2\right] + c^2 dt^2, \tag{2.1.4}$$

where c stands for the speed of light in vacuum for an inertial observer, and (x,y,z) are the space coordinates while t is the temporal coordinate. It is evident that the metric may be positive, negative or null. If $ds^2 = 0$, we say that the metric defines a null displacement, characteristic of light rays.

Riemann spaces are spaces that admit a metric of the type (2.1.2) where the covariant metric tensor is symmetric ($g_{ij} = g_{ji}$).

Contraviant Metric Tensor

On considering g_{ij} like a square matrix, we can define a new tensor g^{ij} from the matrix above by the relation:

$$g^{ij} = \frac{\Delta^{ij}}{g}, \tag{2.1.4}$$

where

Δ^{ij} = cofactor matrix for g_{ij}

g = determinant of g_{ij}.

Taken that g_{ij} is symmetric, and from the definition of a matrix inverse, we find that:

$$g_{ij} g^{jk} = \delta_i^k. \tag{2.1.5}$$

The tensor g^{jk} is sometimes called the conjugate tensor of g_{jk}.

In Riemannian spaces, distinct from amorphous spaces, there is no distinction among the covariant tensors, the contravariants or the mixed, when they refer to the same geometric entity. In fact, given any tensor T_k^{ij} we associate to it the tensor

$$S^{ijk} = g^{lk} T_l^{ij}. \tag{2.1.6}$$

Now, let us associate to the tensor S^{ijk} a new tensor:

$$R_k^{ij} = g_{lk} S^{ijl}. \tag{2.1.7}$$

A close examination reveals that

$$R_k^{ij} \equiv T_k^{ij}. \tag{2.1.8}$$

We conclude, therefore, that the metric tensors work as operators that lower down an index or raise up an index. When we apply this operator once there is no new tensor being created; however we have another representation of the same geometrical entity. In order to ensure proper mathematical consistency we shall use the same letter to represent the same entity with some raised or lowered indices, for example T_k^{ij} and T^{ijk}.

From relation (2.1.5) it is evident that Kronecker's delta is the mixed form of the metric tensor.

2.2. Christoffel Symbols

We define now Christoffel symbols of the 1st and 2nd kind, respectively:

$$\Gamma_{ijk} = \frac{1}{2}\left(\frac{\partial g_{ik}}{\partial x^j} + \frac{\partial g_{kj}}{\partial x^i} - \frac{\partial g_{ij}}{\partial x^k}\right), \tag{2.2.1}$$

$$\Gamma^i_{jk} = g^{il}\Gamma_{jkl}. \tag{2.2.2}$$

Notwithstanding the fact that the symbols Γ_{ijk} and Γ^i_{jk} contain indices, they do not represent tensors as we shall show below.

We shall now show that in a change of coordinates from system S to system S' the Christoffel symbol of the 1st kind Γ_{ijk} transforms itself into Γ'_{ijk} given by the following expression:

$$\Gamma'_{ijk} = \Gamma_{lmn}\frac{\partial x^l}{\partial x'^i}\frac{\partial x^m}{\partial x'^j}\frac{\partial x^n}{\partial x'^k} + g_{lm}\frac{\partial x^l}{\partial x'^k}\frac{\partial^2 x^m}{\partial x'^i \partial x'^j}. \tag{2.2.3}$$

Before we derive this relation, we must point out that it is not in the form of a tensor transformation: the right hand side of the above formula contains two terms summed, the 1st of which if if stood alone, would warrant the tensorial nature of the Christoffel symbol of the 1st kind and consequently, of the 2nd kind too.

Demonstration:

From the definition property of a covariant 2nd rank tensor applied to the metric tensor:

$$g'_{jk} = \frac{\partial x^l}{\partial x'^j}\frac{\partial x^m}{\partial x'^k} g_{lm}. \tag{2.2.4}$$

Applying the partial derivative symbol relative to x'^n to both sides of the above equation, we find by the usual Differential Calculus rules:

$$\frac{\partial g'_{jk}}{\partial x'^n} = \frac{\partial x^l}{\partial x'^j}\frac{\partial x^m}{\partial x'^k}\frac{\partial x^p}{\partial x'^n}\frac{\partial g_{lm}}{\partial x^p} + \frac{\partial^2 x^l}{\partial x'^n \partial x'^j}\frac{\partial x^m}{\partial x'^k} g_{lm} + \frac{\partial^2 x^m}{\partial x'^n \partial x'^k} g_{lm}\frac{\partial x^l}{\partial x'^j}. \tag{2.2.5}$$

By applying repeatedly relation (2.2.5) to the definition of Γ'_{ijk}, i.e., the same definition in the coordinate system S', and remembering the definition of the 1st kind symbols in the system S we obtain result (2.2.3) straightforwardly.

We could equally well obtain the transformation law for the gamma symbols of the 2nd kind and find:

$$\Gamma'^i_{jk} = \Gamma^l_{mn}\frac{\partial x'^i}{\partial x^l}\frac{\partial x^m}{\partial x'^j}\frac{\partial x^n}{\partial x'^k} + \frac{\partial x'^i}{\partial x^n}\frac{\partial^2 x^n}{\partial x'^j \partial x'^k}. \tag{2.2.6}$$

We remark that in the above expression, the first rhs term is the one that, on standing alone would guarantee the tensor character of the gammas, which is not the case here because of the 2nd term in the rhs.

2.3. Covariant Derivative

Except for the case of the derivative of an invariant, in which case we obtain a gradient (a covariant vector) or other partial derivatives of tensors do not result in other tensors. Let us show an example of the above with a contravariant vector V^i. By definition: $V'^i = \frac{\partial x'^i}{\partial x^j} V^j$.

On taking partial derivatives, say relative to x'^k with find:

$$\frac{\partial V'^i}{\partial x'^k} = \frac{\partial x'^i}{\partial x^j}\frac{\partial V^j}{\partial'^k} + \frac{\partial^2 x'^i}{\partial x'^k \partial x^j} V^j = \frac{\partial x'^i}{\partial x^j}\frac{\partial x^l}{\partial x'^k}\frac{\partial V^j}{\partial x^l} + \frac{\partial^2 x'^i}{\partial x^l \partial x^j}\frac{\partial x^l}{\partial x'^k} V^j. \tag{2.3.1}$$

The first term of the right hand side of (2.3.1) should be the unique to exist if the quantity in the left hand side was a tensor, in this case a second rank mixed tensor. The existence of second derivatives can nevertheless be used to define another kind of derivative, called "covariant derivative" that we shall represent by $\frac{DV^i}{Dx^k}$: it will be equal to the simple partial derivative $\frac{\partial V^i}{\partial x^k}$ plus a corrective term such that the sum will be a tensor.

In order to get the said definition, we first notice that from the law transformation of the Christoffel symbols (2.2.6) the following relation is valid:

$$\frac{\partial^2 x'^i}{\partial x^j \partial x^j} = \Gamma^k_{jl}\frac{\partial x'^i}{\partial x^k} - \frac{\partial x'^m}{\partial x^j}\frac{\partial x'^k}{\partial x^l}\Gamma'^i_{mk}. \tag{2.3.2}$$

We now plug the above result into (2.2.7) and we obtain:

$$\left[\frac{\partial V'^i}{\partial x'^k} + \Gamma'^i_{kl} V'^l\right] = \frac{\partial x'^i}{\partial x^j}\frac{\partial x^l}{\partial x'^k}\left[\frac{\partial V^j}{\partial x^l} + \Gamma^j_{ls} V^s\right]. \tag{2.3.3}$$

We define thus, the covariant derivative of the vector V^j relative to x^l, as:

$$\frac{DV^j}{Dx^l} \equiv \frac{\partial V^j}{\partial x^l} + \Gamma^j_{ls} V^s. \tag{2.3.4}$$

Other usual notations are:

$$\frac{DV^j}{Dx^l} \equiv \nabla_l V^j \equiv V^j_{;l},$$

while the partial derivatives appear with a comma, like

$$\frac{\partial V^j}{\partial x^l} \equiv V^j_{,l}.$$

By means of (2.3.2), we can immediately obtain a definition for the covariant derivative of a covariant vector:

$$V_{j;k} = \frac{\partial V_j}{\partial x^k} - \Gamma^l_{jk} V_l. \tag{2.3.5}$$

In order to obtain covariant derivatives for 2nd rank tensors it suffices to imagine how would be the covariant derivative of the product of two vectors, equal to the given tensor: then we apply the formulae (2.3.4) and (2.3.5) obtaining straightforwardly the result:

$$T^{ij}_{;k} = \frac{\partial T^{ij}}{\partial x^k} + \Gamma^i_{kl} T^{lj} + \Gamma^j_{kl} T^{il} \tag{2.3.6}$$

$$T_{ij;k} = \frac{\partial T_{ij}}{\partial x^k} - \Gamma^l_{ki} T^l_{lj} - \Gamma^l_{kj} T_{il} \tag{2.3.7}$$

$$T^i_{j;k} = \frac{\partial T^i_j}{\partial x^k} + \Gamma^i_{kl} T^l_j - \Gamma^l_{kj} T^i_l. \tag{2.3.8}$$

Higher order covariant derivatives and covariant derivatives of higher rank tensors, can be obtained by extending the above techniques. We can easily infer general rules for those cases which are left to the reader. We also leave up to the reader to show that the usual rules for derivatives of sums and products are the same in the case of covariant derivatives.

We now show one of the most important results of Riemannian Geometry, *verbi gratia*, that the covariant derivative of the metric tensor is always null. We ask the reader to combine relations (2.3.6), (2.3.7), (2.2.1) and (2.2.2) in order to prove that:

$$g_{ij;k} = 0, \text{ and,} \tag{2.3.9}$$

$$g^{ij}_{;k} = 0. \tag{2.3.10}$$

2.4. Curvature Tensors

For simple derivatives, the following commutation relation is valid:

$$\frac{\partial}{\partial x^\alpha}\frac{\partial}{\partial x^\beta} = \frac{\partial}{\partial x^\beta}\frac{\partial}{\partial x^\alpha}.$$

We should not expect however that the covariant derivative can commute in the same way as above in the general case of a Riemannian space:

$$V^i_{;j;k} \neq V^i_{;k;j}.$$

Its very easy to show that:

$$V^i_{;j;k} - V^i_{;k;j} = -R^i_{lkj} V^l \tag{2.4.1}$$

and

$$V_{i;j;k} - V_{i;k;j} = R^l_{ijk} V_l \tag{2.4.2}$$

where

$$R^l_{ijk} \equiv \frac{\partial \Gamma^l_{ik}}{\partial x^j} - \frac{\partial \Gamma^l_{ji}}{\partial x^k} + \Gamma^m_{ik}\Gamma^l_{jm} - \Gamma^m_{ij}\Gamma^l_{km}. \tag{2.4.3}$$

The proof of the above is direct: one has only to apply relations (2.3.6) and (2.3.7), combined with (2.3.4) and (2.3.5).

The fourth rank tensor R^i_{jkl} is the famous curvature tensor or Riemann-Christoffel tensor.

Ricci Tensor

By definition, the Ricci Tensor R_{ij} is given by:

$$R_{ij} = R^l_{ilj}. \tag{2.4.4}$$

Ricci Scalar Curvature

By definition, the Ricci Scalar Curvature R is given by:

$$R \equiv R^i_i \equiv g^{ij} R_{ij}. \tag{2.4.5}$$

Einstein's Tensor

By definition, Einstein's Tensor G_{ij} is given by:

$$G_{ij} = R_{ij} - \tfrac{1}{2} g_{ij} R. \tag{2.4.6}$$

Geodesic Coordinate System

A coordinate system is called geodesic, if in a considered point of spacetime, all the components of the Christoffel symbols are null.

Theorem:
"It is always possible, in a given point in spacetime, to build a geodesic coordinate system".

We now demonstrate this theorem:
Consider a coordinate system S in which, for a given spacetime point the gammas are different from zero: $\Gamma^i_{jk} \neq 0$.
We now go to another system of coordinates S' which has the following property:

$$x'^i = x^i - x^i_0 + \tfrac{1}{2}\left(\Gamma^i_{jk}\right)_0 \left[x^j - x^j_0\right]\left[x^k - x^k_0\right]. \tag{2.4.7}$$

The index zero indicates the values of the quantities in the point where the geodesic coordinate system is defined.
We now obtain the null $\left(\Gamma'_{ijk}\right)_0$ by means of relation (2.2.6) if we introduce the following results obtained from (2.4.7):

$$\left(\frac{\partial x'^i}{\partial x^j}\right)_0 = \delta^i_j$$

$$\left(\frac{\partial x^i}{\partial x'^j}\right)_0 = \delta^i_j$$

$$\left(\frac{\partial^2 x^i}{\partial x'^j \partial x'^k}\right)_0 = \left(-\Gamma^i_{jk}\right)_0.$$

Q.E.D.

Theorem:
"In a geodesic system, the partial derivatives of the metric tensor are all null".

Demonstration

We know that the covariant derivatives of the metric tensor are null; however in a geodesic system of coordinates the covariant derivative of the metric tensor equals the partial derivative, so these are also null.

Bianchi Identity

In a geodesic system of coordinates, all the Christoffel symbols are null, but not necessarily the derivatives are null too. In such case we may write:

$$R^{i}_{jkl;m} = \frac{\partial^2 \Gamma^{i}_{jl}}{\partial x^m \partial x^k} - \frac{\partial^2 \Gamma^{i}_{jk}}{\partial x^m \partial x^l}. \tag{2.4.8}$$

On applying the above equation three times, we obtain:

$$R^{i}_{jkl;m} + R^{i}_{jlm;k} + R^{i}_{jmk;l} = 0. \tag{2.4.9}$$

This is a tensorial equation valid for a geodesic coordinate system. However, according to the Fundamental Theorem of Tensor Calculus, derived in Chapter 1, it must be valid in any system: it is called Bianchi Identity.

We now go ahead to perform the most important calculation involving Einstein's tensor: we show that the covariant divergence of the Einstein Tensor is null:

$$G^{i}_{j;i} = 0.$$

Before doing the calculation, one word of caution is necessary: when speaking about the divergence of a tensor, in Riemannian space, the covariant divergence is implied always. Recall now Bianchi identity (2.4.8), which we now cast in the form:

$$R_{pjkl;m} + R_{pjmk;l} + R_{pjlm;k} = 0. \tag{2.4.10}$$

The last equation is obtained from (2.4.8) by multiplying both sides of the equation by g_{pi} and remembering that the covariant derivative of the metric tensor is zero. On the same token, we now multiply by g^{pk} and obtain, given the definition of the Ricci tensor (2.4.4) that:

$$R_{jl;m} - R_{jm;l} + R^{k}_{jlm;k} = 0.$$

We now contract indices by multiplication with g^{jl} and get:

$$R_{;m} - R^{l}_{m;l} - R^{k}_{m;k} = 0.$$

The last two terms are equal because the indices "l" and "k" are dummies. We then find:

$$\left(R^{k}_{m} - \tfrac{1}{2}\delta^{k}_{m} R\right)_{;k} = 0.$$

Q.E.D

Now we shall make our first encounter with the cosmological constant (lambda): on remembering that the covariant derivative of a scalar like R, is equal to the usual derivative, we may write with impunity the following relation which is equivalent to the above:

$$\left[R^k_m - \tfrac{1}{2}\delta^k_m \left(R - 2\Lambda\right)\right]_{;k} = 0. \qquad (2.4.11)$$

Albert Einstein was led to introduce the lambda constant into the Theory of General Relativity as a mathematical artifact which allowed him to obtain, with a proper adjustment of the value of lambda, a cosmological solution for the Universe that turned it static. We shall further study the consequences of the introduction of the "cosmological" constant in later Chapters.

References for Further Study

Same as in last Chapter. See also Raychaudhuri et al (1992) and Ray d'Inverno (1993).

Part II

Introduction to General Relativity

Chapter 3

Basic Theory

3.1. The Basic Principles

Should a theory be "beautiful" in order to be acceptable? — In my subjective opinion, YES! For women, intelligence does the job. South American judges and politicians are usually clever; they get rich, but they are seldom intelligent. Some are even crazy. Also, some scientists (but not me!). Let the ugly ones, forgive me, because for theories, beauty is fundamental.

General Relativity Theory is an intelligent theory of Gravitation, endowed with beautiful equations. It acknowledges that, in a given point of space, an accelerated system is equivalent to a local inertial one, subject to a gravitational field (Principle of Equivalence). This principle has experimental basis in Eötvös experiment, which suggests the equality between inertial and gravitational mass. In Newtonian Mechanics the origin of the inertial mass comes from the second Newton's Law (forces are proportional to accelerations; the constant of proportionality is inertial mass); gravitational mass, is the one which makes the weight of a body proportional to the acceleration of gravity (the constant of proportionality being the gravitational mass).

GRT obeys the covariance principle for the laws of Physics, as already commented. The father of the theory, Albert Einstein, tried to find equations for the motion of bodies in the presence of gravitational fields, which he expected to reduce to the known laws of Special Relativity, when the gravitational field would be switched off; and in the case of weak gravitational fields, and low speeds, when compared to the speed of light, Einstein expected that the GRT equations reduced to those of Newtonian gravitation.

We shall see later that a particle's acceleration in a gravitational field, is proportional to the Christoffel symbols. As it was showed in last chapter, it is always possible, by means of a coordinate transformation (to a geodesic system), to turn to zero the Christoffel symbols at a given point of spacetime. Then, in the new localized coordinate system, at that point, a body feels no acceleration: this makes sense with the Equivalence Principle. For more on "principles", we refer to Chapter 6.

3.2. Energy-Momentum Tensor

For a continuum medium, it can be shown within the Special Relativity Theory, which is the valid theory in the absence of gravitation, that there exists a symmetric tensor, T'^{ij}, so that:

$$\frac{\partial T'^{ij}}{\partial x^i} = D^j, \tag{3.2.1}$$

where D^j stands for the external quadriforce acting on the considered system. If the system is free from external forces, – or if it is not, and then, we include the external environment into the old system, so that the prior external forces become now internal, – we know that we can find a new tensor T^{ij} called the energy momentum tensor of the system, so that its usual divergence is null:

$$\frac{\partial T^{ij}}{\partial x^i} = 0. \tag{3.2.2}$$

The above equation works for a isolated system, and we shall show below that, for a perfect fluid, the (3.2.2) equations reduce themselves to the known continuity and Euler equations of Newtonian Mechanics. For a perfect fluid, we find tentatively an expression for the energy tensor that obeys what we said above. We now offer the final result,

$$T^{ij} = \left[\left(\rho + \tfrac{p}{c^2}\right) u^i u^j - g^{ij} \tfrac{p}{c^2}\right], \tag{3.2.3}$$

where ρ is the energy density of the fluid. p is the pressure and u^i the quadrivelocity of the fluid, defined by

$$u^i = \frac{dx^i}{ds},$$

where c is the speed of light in vacuum for an inertial observer, and the ds^2 is the metric of Special Relativity:

$$ds^2 = -dx^\nu dx^\nu + \left(dx^0\right)^2 = -dx^\nu dx^\nu + c^2 dt^2. \tag{3.2.4}$$

Notice that Greek indices are used here, for three dimensional spaces; for instance ν =1,2,3. When we reduce to Newtonian Gravitation we must obey the following limits:

$$\begin{aligned} u^\nu &\longrightarrow \frac{v^\nu}{c} = \frac{dx^\nu}{cdt} \\ u^0 &\longrightarrow 1 \\ c &\longrightarrow \infty \\ t &\equiv t'. \end{aligned} \tag{3.2.5}$$

For more details, consult a Special Relativity book like Synge (1965). Now we show that our choice for T^{ij} as given by (3.2.3) for a perfect fluid, reduce to the well known continuity and Euler equations.

1st) Continuity Equation.

From (3.2.2), given the symmetry of T^{ij} we obtain:

$$\frac{\partial T^{0i}}{\partial x^i} = 0. \tag{3.2.6}$$

From the above equation we obtain:

$$\frac{\partial}{\partial x^i}\left[\left(\rho + \frac{p}{c^2}\right) u^0 u^i - g^{0i}\frac{p}{c^2}\right] = 0.$$

From (3.2.4) we find $g^{0i} = g^{00} = 1 \quad \text{if} \quad i = 0.$
We also find that $g^{0i} = 0$ if $i \neq 0$.
Therefore we shall find

$$\frac{\partial}{\partial x^\nu}\left[\left(\rho + \frac{p}{c^2}\right) u^0 u^\nu - \frac{p}{c^2}\right] + \frac{\partial}{\partial x^0}\left[\left(\rho + \frac{p}{c^2} - \frac{p}{c^2}\right)\right] = 0.$$

In the limit (3.2.5) we obtain:

$$\frac{\partial}{\partial x^\nu}\left(\rho \frac{v^\nu}{c}\right) + \frac{1}{c}\frac{\partial \rho}{\partial t} = 0,$$

which can also be simplified into:

$$\frac{\partial}{\partial x^\nu}\left(\rho v^\nu\right) + \frac{\partial \rho}{\partial t} = 0,$$

which is the famous continuity equation of fluid dynamics in the Newtonian case.

2nd) Euler's equation.

We obtained (3.2.6) by making $j = 0$ in (3.2.2). Likewise, we now make

$$j = i = 0, 1, 2, 3, \text{ or } \frac{\partial T^{i\nu}}{\partial x^i} = 0. \tag{3.2.7}$$

We find explicitly:

$$\frac{\partial}{\partial x^\nu}\left[\left(\rho + \frac{p}{c^2}\right) u^\nu u^\mu - g^{\mu\nu}\frac{p}{c^2}\right] + \frac{\partial}{\partial x^0}\left[\left(\rho + \frac{p}{c^2}\right) u^\nu u^0 - g^{\mu 0}\frac{p}{c^2}\right] = 0.$$

Going to the desired limit:

$$\frac{\partial}{\partial x^\nu}\left[\rho \frac{v^\nu v^\mu}{c^2} - g^{\mu\nu}\frac{p}{c^2}\right] + \frac{1}{c}\frac{\partial}{\partial t}\left(\rho \frac{v^\mu}{c^2}\right) = 0.$$

We now multiply all the terms by c^2, and taking to account that $g^{\mu\nu}$ reduces to the delta of Kronecker - $\delta^{\mu\nu}$ we obtain:

$$\frac{\partial}{\partial t}\left(\rho v^\mu\right) + \frac{\partial}{\partial x^\nu}\left[\rho v^\nu v^\nu + \delta^{\mu\nu} p\right] = 0. \tag{3.2.8}$$

This is Euler's equation which is essentially the fluid dynamical second Newton's Law as we shall now show.

Newton's second Law for perfect fluids can be written:

$$\frac{d}{dt}(\rho\vec{v}) = -grad\ p = -\nabla p. \tag{3.2.9}$$

From the Differential Calculus of many variables we find that:

$$\frac{d}{dt}(\rho\vec{v}) \equiv \frac{\partial}{\partial t}(\rho\vec{v}) + \frac{\partial}{\partial x^\nu}(\rho\vec{v})v^\nu. \tag{3.2.10}$$

Now

$$v^\nu = \frac{dx^\nu}{dt}$$

(This is Newtonian definition of velocity).

We now rewrite (3.2.9) by means of (3.2.10) in a particular generic component v^μ of velocity and we retrieve equation (3.2.8) QED.

3.3. T^{ij} for the Electromagnetic Field

Maxwell's equations for the electromagnetic field are given by

$$\begin{aligned} \text{div}\ \vec{E} &= \rho_e, \\ \text{rot}\ \vec{B} &= \frac{\partial\vec{E}}{\partial t} + \vec{J}, \\ \text{div}\ \vec{B} &= 0, \\ \text{rot}\ \vec{E} &= -\frac{\partial\vec{B}}{\partial t}, \end{aligned} \tag{3.3.1}$$

where $\vec{E}$ e $\vec{B}$ stands for the electric and magnetic fields, $\vec{J}$ and ρ_e are the densities of electric current and charge. College students who find strange the above equations may benefit from Purcell (1985), Chapter 9.

We define now the Electromagnetic field tensor F^{ij} by giving its components in terms of the above fields:

$$\begin{aligned} F^{12} &= B_3 \\ F^{01} &= E_1 \\ F^{23} &= B_1 \\ F^{02} &= E_2 \\ F^{31} &= B_2 \\ F^{03} &= E_3 \\ F^{ij} &= -F^{ji}. \end{aligned}$$

The reader may easily demonstrate that Maxwell's equations in terms of the antisymmetric tensor F^{ij} become:

$$\frac{\partial}{\partial x^i} F^{ij} = -J^j, \tag{3.3.2a}$$

and

$$\frac{\partial}{\partial x^i} F_{jk} + \frac{\partial}{\partial x^j} F_{ki} + \frac{\partial}{\partial x^k} F_{ij} = 0, \tag{3.3.2b}$$

where

$$J^0 = \rho_e.$$

Textbooks in Special Relativity usually demonstrate that, for the electromagnetic field, the pure energy-momentum tensor, as we defined earlier, is endowed with property (3.2.2), namely, that the divergence of the energy momentum tensor is null, and that it is given by:

$$T^{ij} = F^i_k F^{jk} + \frac{1}{4} g^{ij} F_{kl} F^{kl}. \tag{3.3.3}$$

For details on this and other matters, consult Synge (1965) or references in Chapter 1.

A special comment must be made with reference to the electromagnetic energy tensor field: we defined it in terms of its components in the reference system where Maxwell's equations were cast; it is an experimental result, that the physical quantities represented by it, change from one coordinate system into another, in such a way that electromagnetic energy tensor field is really deserving the word "tensor",.i.e. for example, relations (3.3.3) and (3.2.2), are true in any coordinate system.

3.4. Einstein's Field Equations

Before going to Einstein's field equations, we must remember that the obvious generalization of property (3.2.2) which is valid for Minkowski's spacetime, would be to introduce covariant derivatives instead of simple derivatives when gravitation is present and Riemannian geometry takes place. So reasoned Albert Einstein, and the key idea was that the field equations of General Relativity should reduce to the equations of Special Relativity in the absence of gravitation. All equations should be written in tensor form and for weak gravitational fields they should reduce to Newtonian gravity laws. So, in the presence of gravitational fields, the covariant divergence of the energy momentum tensor, should be conserved, in the covariant derivative sense:

$$T^{ij}_{;j} = 0. \tag{3.4.1}$$

If the equation for matter fields, is written with the tensor T^{ij} obeying the above property and on remembering that in Riemannian geometry which has a similar law of conservation for the Einstein's tensor, namely

$$G^{ij}_{;j} = 0, \tag{3.4.2}$$

which we studied in Chapter 2, it would be reasonable that matter and geometry of space were tied by the equations

$$\boxed{G^{ij} = -\kappa T^{ij}} \qquad \text{(Einstein's field Equations)} \tag{3.4.3}$$

because then the properties (3.4.2) and (3.4.1) would be automatically fulfilled: we may say that matter makes the geometry. Precisely speaking, if we know the energy density and the pressure of a fluid, we may build its energy momentum tensor, and then we shall know how this matter curves spacetime through the Einstein tensor, which reflects the properties of the curvature of spacetime. The constant κ is called Einstein's gravitational constant and it is closely related to Newtonian gravitational constant G as we shall show later on. The plausibility for (3.4.3), lies in the paragraph following formula (3.4.5) below.

Einstein's field equations should reduce for weak gravitational fields and low speeds when compared with the speed of light, to the Newtonian equations of Poisson:

$$\frac{\partial^2 \Psi}{\partial x^\nu \partial x^\nu} = -4\pi G\rho \tag{3.4.4}$$

where Ψ is the Newtonian scalar potential, G is the Newton's gravitational constant and ρ stands for mass density, while the potential Ψ relates with Newton's second law by

$$\frac{d^2 x^\nu}{dt^2} = -\frac{\partial \Psi}{\partial x^\nu}. \tag{3.4.5}$$

In order that this last expression is also verified we need an additional hypothesis: a particle subject to a gravitational field moves according to a timelike geodesic of Riemann's spacetime. Roughly speaking, we can say that we have found the following reduction, which yields (3.4.4) from (3.4.3):

$$\boxed{G^{ij} \to \frac{\partial^2 \Psi}{\partial x^\nu \partial x^\nu}} \quad \text{while} \quad \boxed{T^{ij} \to -\tfrac{1}{2}\kappa\rho}.$$

In Chapter 5, section 5.2, we shall see that the equation of geodesics expresses the fact that the tangent vector to the curve stays parallel to itself during motion.

The equation derived for geodesics in Riemann spaces is

$$\frac{d^2 x^i}{ds^2} + \Gamma^i_{jk} u^j u^k = 0. \tag{3.4.6}$$

Now, we shall show that geodesics reduce to straight lines of Euclidean geometry in the absence of gravitation. Indeed, if gravitation is absent we can choose Cartesian coordinates where the Christoffel symbols are null. We are left with the equation:

$$\frac{d^2 x^i}{ds^2} = 0.$$

For $i = 0$ we find

$$\frac{d^2 x^0}{ds^2} = 0$$

which has the solution $x^0 = s$. With this solution we find:

$$\frac{d^2x^i}{dx_0^2} = 0 \quad \text{or}$$

$$\frac{d^2x^i}{dt^2} = 0$$

which are the equations of a straight line.

QED.

We remember that straight lines are the shortest curves between two points in Euclidean Geometry and that a similar role can be attached to geodesics in the Riemannian geometry (stationary property).

Einstein proposed, in the year 1916, that the geodesic postulate should be incorporated to GRT as a separate one. Eleven years later however, Einstein and Grommer showed that this postulate could be indeed demonstrated with help of the field equations alone. Nevertheless, there are some restrictions for the derivation of this postulate from the field equations; for instance, a spinning particle does not follow geodesics. (see details in Papapetrou, 1974 – Section 42).

In this Chapter, we shall offer a particular case ($p = 0$ matter, also called "dust"), where we shall show how to derive the equation of geodesics from field equations (3.4.3).

3.5. Introducing the Cosmological Constant

The Einstein tensor, G^{ij}, as we recall from Chapter 2, had the property that its covariant divergence was null. Because any tensor proportional to the metric tensor has zero covariant derivative we may say that any tensor of the type:

$$E^{ij} = G^{ij} + \Lambda g^{ij}, \tag{3.5.1}$$

where Λ is an arbitrary constant, has also a null covariant divergence:

$$E^{ij}_{;j} = 0. \tag{3.5.2}$$

Albert Einstein furthermore concluded that his field equations should be conceived with the lambda constant, employing thus the tensor E^{ij} instead of G^{ij} so that we would find:

$$\boxed{E^{ij} = -\kappa T^{ij}} \tag{3.5.3}$$

We call the above field equations (3.5.3) as the field equations with a cosmological constant. The word "cosmological" is not without reason: for local Physics experiments we do not need such constant. It is only when large portions of the Universe are considered, that lambda is really needed. However, even in other fields, lambda may be important, as the energy density of the vacuum, as stated below.

We leave as an exercise to the reader to follow all the steps in order to demonstrate that, alternatively, for a perfect fluid energy-momentum tensor T^{ij} as given by (3.2.3), we can define a new energy-momentum tensor

$$\widetilde{T}^{ij} = \left(\widetilde{\rho} + \tfrac{\widetilde{p}}{c^2}\right) u^i u^j - g^{ij} \tfrac{\widetilde{p}}{c^2} \tag{3.5.4}$$

where there is a redefinition of energy density, and fluid pressure, of the type:

$$\tfrac{\widetilde{p}}{c^2} = \tfrac{p}{c^2} - \tfrac{\Lambda}{\kappa}; \tag{3.5.5}$$

$$\widetilde{\rho} = \rho + \tfrac{\Lambda}{\kappa}.$$

With this new energy-momentum tensor we can write:

$$G^{ij} = -\kappa \widetilde{T}^{ij}. \tag{3.5.6}$$

If the student finds this boring or confusing, be aware that we are dealing with a great problem in GRT: the cosmological constant problem. With the field equations (3.5.6) we have transferred the responsibility for keeping the account of the cosmological term from the Einstein's tensor E^{ij} to a modified energy-momentum tensor $\widetilde{T}^{ij}$ in such a way that we may imagine to have augmented the energy density by a parcel $\frac{\Lambda}{\kappa}$: the question remains, where does this energy come from? The Russian cosmologist Yakov Zel'dovich, in the year 1974, prompted an answer to this question: the additional energy comes - yes!!! - from the energy density of the vacuum. So, the zero field energy of Quantum Mechanics can find its way into General Relativity, which is a Classical Theory, through the lambda constant. Zel'dovich and others have hinted that besides the normal fluid for matter, the Universe is permeated with a second fluid which is offered to us "free of charge" and contributes with an energy density given by $\frac{\Lambda}{\kappa}$ and a negative pressure, $-\frac{\Lambda}{\kappa}$ as can be checked from the first relation numbered (3.5.5). Physicists today joke that Zel'dovich cleaned the Geometry from an annoying lambda and threw away the burden to the Physics part of the field equations, in the r.h.s.

New cosmological discoveries in the beginning of the twenty-first century showed that the Universe has been accelerating its expansion, which points to a tiny but non zero cosmological constant to pervade the equations of cosmology. Albert Einstein was led in some period of his life to regret having introduced the lambda constant, describing this fact as the greatest blunder in his life. Despite the fact that he never wrote such a thing, his contemporary colleague George Gamow recalled in his memories having heard Einstein complain in that way, about the lambda constant. It is refreshing to mention that Particle Physicists require that, in the very early Universe, the energy of the vacuum be extraordinarily huge. Nevertheless, we shall see in next Chapter that the Schwarzschild's solution for the solar system, while neglecting the cosmological term, is a very accurate description of astronomical observations near the Sun and the Earth, thus pointing to a very small lambda. In fact, it should be 10^{-120} orders of magnitude lower than what Particle Physics needs for the very early Universe. We shall see in further chapters, that some physicists, like Orfeu Bertolami and Marcelo Berman worked with the possibility of a lambda term not constant but decaying with the square of the age of the Universe. Alternatively Chen and Wu and Berman also worked with a solution for the lambda problem, that would make for a lambda term decaying with the square of the scale-factor (sometimes called the "radius") of the Universe. For other possibilities, see Overduin and Cooperstock (1998).

3.6. Reduction of GRT into Newtonian Physics

Our purpose in what follows shall be to prove that the Newtonian potential Ψ is related to the metric tensor, so that g^{ij} are representative of the gravitational potentials in Einstein's theory. Furthermore, it will be shown that the condition for the field equations (3.4.3) to reduce to the Newtonian Poisson equation is that:

$$\kappa = \frac{8\pi G}{c^2}. \tag{3.6.1}$$

Note that the field equations can be written with another sign in the right hand side, if κ is defined as $(-\kappa)$.

1stCondition: Relations betweeng^{ij}andΨ.

Equation (3.4.6), the geodesic equations, do reduce in the static case and for weak gravitational fields, like it follows. A weak field differs very little from Minkowskian metric. Let us write

$$g_{ij} = \eta_{ij} + \gamma_{ij}, \tag{3.6.2}$$

where η_{ij} is a the diagonal matrix $(+1,-1,-1,-1)$ and the γ_{ij}'s are very less than unity in absolute value. If we now impose a static metric, i.e.,

$$\frac{\partial g_{ij}}{\partial x^0} \approxeq 0,$$

and if we consider low speeds in comparison with the speed of light, the (3.4.6) should reduce to

$$\frac{d^2x^i}{ds^2} + \Gamma^i_{00} \approxeq 0,$$

where

$$\Gamma^i_{00} = \frac{1}{2} g^{ij} \left(\frac{\partial g_{0j}}{\partial x^0} + \frac{\partial g_{0j}}{\partial x^0} - \frac{\partial g_{00}}{\partial x^j} \right) \approxeq -\frac{1}{2} g^{ii} \frac{\partial g_{00}}{\partial x^i} \approxeq -\frac{1}{2} \frac{\partial \gamma_{00}}{\partial x^i} \approxeq -\frac{1}{2} \frac{\partial g_{00}}{\partial x^i}.$$

Then

$$\frac{d^2x^i}{dt^2} \approxeq -\frac{c^2}{2} \frac{\partial g_{00}}{\partial x^i}. \tag{3.6.3}$$

On comparing (3.6.3) with the Newtonian equation:

$$\frac{d^2x^i}{dt^2} = -\frac{\partial \Psi}{\partial x^i},$$

we find that:

$$\Psi \approxeq \frac{g_{00}c^2}{2} + \text{constant}.$$

In order to determine the value of the constant above we shall consider Minkowski's metric, which is the limiting metric for points that are infinitely far from the masses. In this case:

$$g_{00} \rightarrow 1,$$

$$\Psi \rightarrow 0,$$

and then

$$\Psi = \frac{c^2}{2} + \text{constant} \rightarrow 0,$$

or,

$$\Psi = \frac{g_{00}c^2}{2} - \frac{c^2}{2},$$

and then

$$g_{00} = 1 + \frac{2\Psi}{c^2}. \tag{3.6.4}$$

Therefore, g_{00} stands for a gravitational potential, and, consequently we justified in calling the metric tensor components as the "gravitational potentials". It is also clear from the intermediary step where we found,

$$\frac{d^2x^i}{ds^2} \approxeq -\Gamma^i_{00},$$

that the Christoffel symbols represent accelerations, as we had advanced earlier.

2nd Condition: Obtainment of the value for Einstein's gravitational constant.

In the same conditions as above we now undertake the calculation of curvature tensor components R^i_{jkl}. As the Christoffel symbols are very small quantities, their products are negligible. So we can write:

$$R^i_{jkl} \approxeq \frac{\partial \Gamma^i_{jl}}{\partial x^k} - \frac{\partial \Gamma^i_{jk}}{\partial x^l}.$$

Therefore, we obtain Ricci tensor by:

$$R_{jk} \approxeq \frac{1}{2}\frac{\partial}{\partial x^k}\left[\frac{\partial \gamma_{ji}}{\partial x^i} + \frac{\partial \gamma_{ii}}{\partial x^j} - \frac{\partial \gamma_{ij}}{\partial x^i}\right] - \frac{1}{2}\frac{\partial}{\partial x^i}\left[\frac{\partial \gamma_{ij}}{\partial x^k} + \frac{\partial \gamma_{ki}}{\partial x^j} - \frac{\partial \gamma_{jk}}{\partial x^i}\right] =$$

$$\frac{1}{2}\left[\frac{\partial^2 \gamma_{ii}}{\partial x^j \partial x^k} + \frac{\partial^2 \gamma_{jk}}{\partial x^i \partial x^i} - \frac{\partial^2 \gamma_{ij}}{\partial x^i \partial x^k} - \frac{\partial^2 \gamma_{ki}}{\partial x^i \partial x^j}\right].$$

In the particular case $j = k = 0$, we obtain:

$$R_{00} \approxeq \frac{1}{2}\left[\frac{\partial^2 \gamma_{ii}}{\partial x^0 \partial x^0} + \frac{\partial^2 \gamma_{00}}{\partial x^i \partial x^i} - 2\frac{\partial^2 \gamma_{10}}{\partial x^i \partial x^0}\right].$$

In the static case,

$$R_{00} \approxeq \frac{1}{2}\frac{\partial^2 \gamma_{00}}{\partial x^i \partial x^i} \approxeq \frac{1}{2}\nabla^2 \gamma_{00} \approxeq \frac{1}{2}\nabla^2 g_{00}.$$

In accordance with (3.6.4) the last expression reduces to:

$$R_{00} \approxeq \frac{1}{c^2}\nabla^2 \Psi. \tag{3.6.5}$$

We now are going to make a detour, and show that Einstein's field equations may be expressed, alternatively as

$$R_{ij} = \kappa\left[\tfrac{1}{2}g_{ij}T - T_{ij}\right], \tag{3.6.6}$$

where

$$T \equiv T_i^i.$$

By contracting the (3.4.3) field equation, we obtain:

$$G_i^i = \kappa T_i^i, \text{ or}$$

$$R = \kappa T. \tag{3.6.7}$$

If we now plug (3.6.7) in equation (3.4.3), we obtain the desired relation (3.6.6).
After this detour, we return to (3.6.5) and we apply (3.6.6):

$$\tfrac{1}{c^2}\nabla^2\Psi = \kappa\left(\tfrac{1}{2}g_{00}T - T_{00}\right). \tag{3.6.8}$$

We now suppose the absence of other fields like for instance the electromagnetic; we suppose altogether that the matter distribution is that of "dust", i.e., $p \cong 0$, or,

$$T^{ij} = \rho u^i u^j.$$

In the static case we shall find:

$$T_{00} \approxeq \rho\,(g_{00})^2 \approxeq \rho,$$

and

$$T \approxeq \rho\,(g_{00})^2 \approxeq \rho.$$

On taking this values into (3.6.8), we obtain,

$$\tfrac{1}{c^2}\nabla^2\Psi \approxeq -\tfrac{\kappa\rho}{2},$$

or

$$\nabla^2\Psi \approxeq -\tfrac{\kappa c^2}{2}\rho.$$

If we compare the above result with Poisson's equation (3.4.4):

$$\nabla^2\Psi = -4\pi G\rho, \tag{3.4.4}$$

we obtain:

$$\boxed{\kappa = \frac{8\pi G}{c^2}} \tag{3.6.1}$$

3.7. The Geodesics Equation - a Derivation

We now derive, for the case of "dust" matter, immersed in a gravitational field, the equation of motion for each particle (each particle will follow a timelike geodesic).

For this case we found above that:

$$T^{ij} = \rho u^i u^j.$$

From the conservation of the covariant divergence of the energy-momentum tensor we have:

$$T^{ij}_{;j} = 0.$$

It follows that:

$$\left(\rho u^j\right)_{;j} u^i + \rho u^j u^i_{;j} = 0. \quad (3.7.1)$$

On multiplying the last equation per u^i, it follows that:

$$\left(\rho u^j\right)_{;j} u^i u_i + \rho u^i_{;j} u_i u^j = 0. \quad (3.7.2)$$

From the definition of quadrivelocities we find:

$$u^2 = g_{ij} u^i u^j = g_{ij} \frac{dx^i}{ds}\frac{dx^j}{ds} = \frac{ds^2}{dsds} = 1.$$

In other words,

$$u_i u^i = 1. \quad (3.7.3)$$

On taking derivatives of both sides with respect to x^j, we find:

$$u_{i;j} u^i + u_i u^i_{;j} = 0,$$

or else,

$$u^i_{;j} u_i = 0.$$

Taking this result to (3.7.2), we find:

$$\left(\rho u^j\right)_{;j} = 0. \quad (3.7.4)$$

Taking this result to (3.7.1), we obtain:

$$u^j u^i_{;j} = 0.$$

From the definition of covariant derivatives we have, for a contravariant vector,

$$u^i_{;j} \equiv \left[\frac{\partial u^i}{\partial x^j} + \Gamma^i_{kj} u^k\right].$$

Therefore,

$$u^i_{;j}u^j \equiv \left[\frac{\partial u^i}{\partial x^j} + \Gamma^i_{kj}u^k\right] u^j = 0.$$

Or,

$$\frac{d^2x^i}{ds^2} + \Gamma^i_{kj}\frac{dx^j}{ds}\frac{dx^k}{ds} = 0. \tag{3.7.5}$$

This is the famous geodesics equation, derived from Einstein's field equations, so that it has not to be adopted by means of an additional postulate within GRT. In the case of a particle not subject to the gravitational field, a straight line is our geodesic in Minkowski's space (Special Relativity) or in the three-dimensional Euclidean space (Newtonian Mechanics).

By employing variational calculus techniques, it is possible to demonstrate effectively that a geodesic equation in Riemannian geometry represents the curve, between two points of longest or shortest path. This is an additional evidence that the GRT reduces to Newtonian Mechanics where the counterpart of a geodesic is the straight line, known to be the shortest distance between two points in Euclidean space.

3.8. Gravitational Radiation

[We follow closely, here, the notation used by Kenyon (1990)].

We shall show that, in empty space, the linearized field equations, yield the wave equation. In empty space Einstein's equations become:

$$G_{\mu\nu} \equiv R_{\mu\nu} = 0 \tag{3.8.1}$$

Let us suppose that there is a small perturbation in the otherwise Minkowski's metric background:

$$g_{\mu\nu} = \eta_{\mu\nu} + h_{\mu\nu}, \tag{3.8.2}$$

where $\eta_{\mu\nu}$ stands for the Minkowski's metric coefficients, and the $h_{\mu\nu}$ are small perturbations. Notice that, the present linearized theory, differs from the previous one in Section 3.6, which is more restrictive.

From the definition of the Christoffel symbols, we obtain:

$$\Gamma^{\alpha}_{\beta\delta} = \frac{1}{2}g^{\alpha\nu}[h_{\beta\nu,\delta} - h_{\delta\beta,\nu} + h_{\nu\delta,\beta}]. \tag{3.8.3}$$

On the other hand, to first order in the perturbation, products of the $h_{\mu\nu}$'s are neglected and we find:

$$R_{\beta\delta} \cong \Gamma^{\alpha}_{\beta\delta,\alpha} - \Gamma^{\alpha}_{\beta\alpha,\delta} \cong \frac{1}{2}g^{\alpha\nu}[h_{\nu\delta,\beta\alpha} - h_{\delta\beta,\nu\alpha} + h_{\alpha\beta,\nu\delta} - h_{\alpha\nu,\beta\delta}] \cong 0. \tag{3.8.4}$$

(the zero result is due to the field equations).

As the $g^{\mu\nu}$'s also raise the subscripts of the $h_{\mu\nu}$'s, we find:

$$h^{\alpha}_{\delta,\beta\alpha} - h_{\delta\beta,\varepsilon\alpha}\eta^{\varepsilon\alpha} - h^{\alpha}_{\alpha,\beta\delta} + h_{\alpha\beta,\beta\delta}\eta^{\beta\alpha} \cong 0. \tag{3.8.5}$$

We may try to find a particular solution for the above equation. Suppose that:

$$h^{\alpha}_{\alpha} = 0, \tag{3.8.6}$$

and,

$$h^{\alpha}_{\delta,\alpha} = h_{\alpha\beta,\gamma}\eta^{\gamma\alpha} = 0. \tag{3.8.7}$$

Then we are left with:

$$h_{\delta\beta,\varepsilon\delta}\eta^{\varepsilon\alpha} \cong 0. \tag{3.8.8}$$

Equation (3.8.6) means that the perturbation is traceless; (3.8.7) means that in our particular solution we shall have zero divergence. Relation (3.8.8) is then:

$$\Box^2 h^{\mu\nu} = 0, \tag{3.8.9}$$

where $\Box^2$ stands for the d'Alembertian, i.e.:

$$\Box^2 \equiv -\eta^{\alpha\beta}\frac{\partial}{\partial x^\alpha}\frac{\partial}{\partial x^\beta} \equiv \frac{\partial^2}{\partial x^2} + \frac{\partial^2}{\partial y^2} + \frac{\partial^2}{\partial z^2} - \frac{1}{c^2}\frac{\partial^2}{\partial t^2} \equiv \nabla^2 - c^{-2}\frac{\partial^2}{\partial t^2}.$$

Equation (3.8.9) is the wave equation solution we were expecting. We make also the additional condition,

$$h_{\alpha 0} = 0, \tag{3.8.10}$$

and then, we can check that there is a freedom for choosing arbitrary only two independent components of $h_{\alpha\beta}$, which will be the two polarization states of the wave. The d'Alembertian equation means that the signal runs with the speed of light. In the Quantum outlook, the waves are carried by "gravitons", which are the equivalent of the photons in the electromagnetic case.

Gravitational waves were never been detected directly; nevertheless, there is indirect evidence of their existence, obtained by Hulse and Taylor who obtained the Nobel Prize for their discovery (see Taylor and Weisberg, 1982; 1989). See Section 4.5 for a possible explanation of the negative results, so far, in attempted detections.

Chapter 4

Schwarzschild's Metric and Classical Experimental Tests

4.1. Spherically Symmetric Metrics

We define a metric with spherical symmetry around the origin, as a metric which is form-invariant according to the group of orthogonal transformations, of the type:

$$\widetilde{X} = AX,$$

where,

$$X \equiv (x, y, z),$$

$$\widetilde{X} \equiv (\widetilde{x}, \widetilde{y}, \widetilde{z}),$$

and

$$AA^T = I,$$

where I represents the identity matrix. (For the reader who needs to know, and does not, what are orthogonal transformations, next chapter has a review on it).

A metric is form-invariant if, in a transformation of coordinates $X \to \widetilde{X}$, we have,

$$g_{ij}(x) = \widetilde{g}_{ij}(\widetilde{x}).$$

Notice that the temporal coordinate is excluded from this spherical symmetry definition.

Form-invariants, for this group of transformations, are the following, up to second order in the coordinate differentials:

$$+x^2 + y^2 + z^2,$$

$$+xdx + ydy + zdz,$$

$$+dx^2 + dy^2 + dz^2.$$

In spherical coordinates (r, θ, ϕ), the above invariants become:

$$+r^2,$$

$$+rdr,$$

$$+d\theta^2+\sin^2\theta d\phi^2.$$

We may also conclude, that the following are invariants: $(r, dr, d\theta^2, +\sin^2\theta d\phi^2)$. In this way, the most general spherically symmetric metric in Riemann four-dimensional space (r, θ, ϕ, t) is:

$$ds^2 = A(r,t)dr^2 + B(r,t)(d\theta^2+\sin^2\theta d\phi^2) + C(r,t)drdt + D(r,t)dt^2, \tag{4.1.1}$$

where the four coefficients A, B, C, D are up to now completely arbitrary.

Now let us remember Minkowski's metric:

$$ds^2 = -dx^\nu dx^\nu + c^2dt^2. \qquad (\nu = 1, 2, 3)$$

In spherical coordinates, Minkowski's metric may therefore be expressed by:

$$ds^2 = -dr^2 - r^2(d\theta^2+\sin^2\theta d\phi^2) + c^2dt^2. \tag{4.1.2}$$

The metric which represents a mass distribution in the origin, should reduce, when the radial coordinate becomes very large $(r \longrightarrow \infty)$, to the Minkowski's metric; (in Cosmology, this may not be alike), so we take:

$$B(r,t) = r^2.$$

In the case of a static gravitational field, the functions $A(r,t)$, $C(r,t)$, and $D(r,t)$ should not depend on the time coordinate t. On the other hand, the time dependence of the metric should be symmetric; i.e., the form of the metric can not be altered by a transformation $t \longrightarrow -t$: we conclude that $C = 0$. We are left with the following metric:

$$ds^2 = -\left[e^{\alpha(r)}dr^2 + r^2(d\theta^2+\sin^2\theta d\phi^2)\right] + e^{\beta(r)}c^2dt^2.$$

We choose in the above expression for the metric the exponentials $e^{\alpha(r)}$ and $e^{\beta(r)}$ so that they remain essentially positive and the signature of the metric should not change, at least for the time being.

We now begin the calculation of the metric tensor, the Christoffel symbols and the curvature and Ricci tensors, while attaching the following correspondence:

$$(x^0, x^1, x^2, x^3) \equiv (t, r, \theta, \phi).$$

1st) Non-null covariant metric tensor components:

$$g_{11} = -e^\alpha,$$

$$g_{22} = -r^2,$$

$$g_{33} = -r^2 \sin^2\theta,$$

$$g_{00} = e^{\beta}c^2,$$

$$g = det(g_{ij}) = -e^{(\alpha+\beta)}c^2r^4\sin^2\theta\ .$$

2nd) Non-null contravariant components of the metric tensor:

$$g^{11} = -\frac{1}{e^{\alpha}} = -e^{-\alpha},$$

$$g^{22} = -\frac{1}{r^2},$$

$$g^{33} = -\frac{1}{r^2\sin^2\theta},$$

$$g^{00} = -\frac{1}{e^{\beta}c^2}.$$

3rd) Non-null Christoffel symbols of the second kind:

$$\Gamma^1_{11} = \frac{\alpha'}{2},$$

$$\Gamma^2_{12} = \Gamma^2_{21} = \frac{1}{r}\ ,$$

$$\Gamma^3_{13} = \Gamma^3_{31} = \frac{1}{r},$$

$$\Gamma^0_{10} = \Gamma^0_{01} = \frac{\beta'}{2},$$

$$\Gamma^1_{22} = -re^{-\alpha},$$

$$\Gamma^3_{23} = \Gamma^3_{32} = \mathrm{cotg}\ \theta,$$

$$\Gamma^1_{33} = -re^{-\alpha}\sin^2\theta\ ,$$

$$\Gamma^2_{33} = -\sin\theta\cos\theta,$$

$$\Gamma^1_{00} = -\frac{1}{2}c^2\beta' e^{(\beta-\alpha)}.$$

(the primes indicate derivatives in relation to the radial coordinate).

4th) Ricci tensor non-null components:

$$R_{11} = \frac{\beta''}{2} + \frac{\beta'^2}{4} - \frac{\alpha'\beta'}{4} - \frac{\alpha'}{r},$$

$$R_{22} = e^{-\alpha}\left[r\frac{\beta'}{2} - \frac{r\alpha'}{2} + 1\right] - 1.$$

$$R_{33} = R_{22}\sin^2\theta.$$

$$R_{00} = c^2 e^{(\beta-\alpha)}\left[-\tfrac{1}{2}\beta'' - \tfrac{1}{4}\beta'^2 + \tfrac{1}{4}\alpha'\beta' - \tfrac{\beta'}{r}\right].$$

Schwarzschild's metric exterior solution

The metric obtained by us, shall be certainly a solution for a static distribution, with spherical symmetry around the origin. For points that are exterior to the mass distribution in the origin, Einstein's field equations reduce to:

$$R_{ij} = 0, \tag{4.1.4}$$

because $T_{ij} = 0$.

According to the above calculation, we are left with the equations:

$$\beta'' + \tfrac{1}{2}\beta'^2 - \tfrac{1}{2}\alpha'\beta' - \tfrac{2}{r}\alpha' = 0, \tag{4.1.5}$$

$$\tfrac{1}{2}\beta' r - \tfrac{1}{2}\alpha' r + 1 = e^{\alpha}, \tag{4.1.6}$$

$$\beta'' + \tfrac{1}{2}\beta'^2 - \tfrac{1}{2}\alpha'\beta' - \tfrac{2}{r}\beta' = 0. \tag{4.1.7}$$

Notice that we were left with three independent equations alone, because in our case, $R_{22} = 0 \rightarrow R_{33} = 0$.

By comparing (4.1.5) and (4.1.7), we obtain:

$$\alpha + \beta = \text{constant}. \tag{4.1.8}$$

The above equality should also be valid in spatial infinity where it should be identified by the values in the Minkowski's metric (4.1.2), and

$$e^{\alpha} = e^{\beta} = 1 \qquad \therefore \qquad \alpha \equiv \beta = 0.$$

The constant in (4.1.8) is obviously zero-valued. And then, we find:

$$\alpha = -\beta \text{ at any place}. \tag{4.1.9}$$

From (4.1.6) we now obtain

$$r\alpha' = 1 - e^{\alpha},$$

or in other words,

$$\frac{dr}{r} = \frac{d\alpha}{1-e^{\alpha}}.$$

When we integrate, the above equation, and we call "m" a constant of integration, we find:

$$e^{\alpha} = \frac{1}{1-\frac{2m}{r}}. \tag{4.1.10}$$

Then

$$e^{\beta} = 1 - \frac{2m}{r}. \tag{4.1.11}$$

We have thus obtained Schwarzschild's metric:

$$ds^2 = \frac{-dr^2}{1-\frac{2m}{r}} - r^2(d\theta^2 + \sin^2\theta d\phi^2) + c^2(1 - \frac{2m}{r})dt^2. \tag{4.1.12}$$

The integration constant "m" can be given now a precise value: we remember that very far from the origin, where the mass M is located, the above metric can be identified with the metric of a weak field whose temporal metric coefficient is given by (3.5.4):

$$g_{00} = 1 + \frac{2\Psi}{c^2}. \tag{3.5.4}$$

We conjecture that the potential Ψ very far from the origin, is given by its Newtonian expression namely :

$$\Psi = -\frac{GM}{r}. \tag{4.1.13}$$

Obviously, we can identify the integration constant "m" with:

$$m = \frac{GM}{c^2}. \tag{4.1.14}$$

There is an apparent singularity of the Schwarzschild's metric for the point:

$$r = 2m = \frac{2GM}{c^2}. \tag{3.1.15}$$

This radial distance is called Schwarzschild's radius, and its value for the Earth, is about 9mm. If the Earth's mass would be entirely concentrated inside a sphere of radius smaller then 9mm, we would have a **black-hole**. By the so called process of gravitational collapse, its radius would become even more smaller than that, though I hope that it should not reduce to an infinitesimally small point of zero radius. Through the Schwarzschild's metric, a white-hole can also be envisaged with properties similar to a fountain of matter and photons surging from an apparent point-like source.

4.2. The Motion of Planets and Perihelion Precession

We shall now try to determine an equation for the orbital motion of a planet around the Sun. On considering valid the metric (4.1.12), we remember that each planet must obey approximately the path of a geodesic of the given space-time.

The equations of geodesics is given by (3.7.5), which we reproduce now:

$$\frac{d^2x^i}{ds^2}+\Gamma^i_{kj}\frac{dx^j}{ds}\frac{dx^k}{ds}=0. \tag{3.7.5}$$

With the values obtained for the Γ^i_{jk}'s in the prior section of this Chapter, we obtain the following equations:

$$\frac{d^2r}{ds^2}+\frac{\alpha'}{2}\left(\frac{dr}{ds}\right)-re^{-\alpha}\left(\frac{d\theta}{ds}\right)^2-r\sin^2\theta e^{-\alpha}\left(\frac{d\phi}{ds}\right)^2+\frac{e^{\beta-\alpha}}{2}\alpha'\left(\frac{dt}{ds}\right)^2=0, \tag{4.2.1}$$

$$\frac{d^2\theta}{ds^2}+\frac{2}{r}\frac{dr}{ds}\frac{d\theta}{ds}-\sin\theta\cos\theta\left(\frac{d\phi}{ds}\right)^2=0, \tag{4.2.2}$$

$$\frac{d^2\phi}{ds^2}+\frac{2}{r}\frac{dr}{ds}\frac{d\theta}{ds}-2\cot\theta\frac{d\theta}{ds}\frac{d\phi}{ds}=0, \tag{4.2.3}$$

$$\frac{d^2t}{ds^2}+\beta'\frac{dr}{ds}\frac{dt}{ds}=0. \tag{4.2.4}$$

Let us begin our study of the motion of the planet supposing that initially $\theta=\frac{\pi}{2}$ and $\frac{d\theta}{ds}=0$. From (4.2.2) it follows that:

$$\frac{d^2\theta}{ds^2}=0 \qquad \therefore \qquad \frac{d\theta}{ds}=\text{constant}=0$$

(because zero was its initial value).

On the same token,

$$\theta=\frac{\pi}{2}=\text{constant}. \tag{4.2.5}$$

The planet motion occurs always on the same plane, and is defined by equations (4.2.1), (4.2.3) and (4.2.4).

Now we turn to a solution of those equations. From the expression of the metric, we obtain an obvious solution:

$$1+e^{\alpha}\left(\frac{dr}{ds}\right)^2+r^2\left(\frac{d\phi}{ds}\right)^2-e^{\beta}\left(\frac{dt}{ds}\right)=0. \tag{4.2.6}$$

Equation (4.2.3) becomes now:

$$\frac{du}{dr}\frac{dr}{ds}+\frac{2}{r}\frac{dr}{ds}u=0,$$

where we define $u\equiv\frac{d\phi}{ds}$.

A trivial solution would be $r=$constant. If we drop it, we are left with:

$$\frac{du}{u}=2\frac{dr}{r},$$

If C is a constant of integration, we find from the above:

$$u = \frac{d\phi}{ds} = \frac{C}{r^2}. \tag{4.2.7}$$

If we now define $n \equiv \frac{dt}{ds}$, equation (4.2.4) becomes:

$$\frac{dn}{d\beta} + n(\beta) = 0,$$

By integration, we find:

$$n = n(\beta) = \frac{dt}{ds} = Le^{-\beta}, \tag{4.2.8}$$

where L is a constant of integration.

We now take (4.2.8) to (4.2.6), and on remembering the values of α and β which we obtained in the prior section, we find:

$$1 + \frac{m}{r} + \left(\frac{dr}{ds}\right)^2 + r^2\left(\frac{d\phi}{ds}\right)^2\left(1 + \frac{m}{r}\right) - L^2 = 0. \tag{4.2.9}$$

Let us define a new constant:

$$h = L^2 - 1.$$

Equation (4.2.9) may now be written as:

$$\left(\frac{dr}{ds}\right)^2 + r^2\left(\frac{d\phi}{ds}\right)^2 + \frac{m}{r}\left[1 + r^2\left(\frac{d\phi}{ds}\right)^2\right] = h, \tag{4.2.10}$$

while from (4.2.7) it follows that:

$$r^2\left(\frac{d\phi}{ds}\right)^2 = C. \tag{4.2.11}$$

Equations (4.2.5), (4.2.10), and (4.2.11) rule the motion of a planet according to GRT. In order to interpret the above solution we turn our eyes to the corresponding solution in Newtonian Mechanics whose three laws are:

1st Law: Central motion is contained on a plane.

$\therefore\ \theta = \frac{\pi}{2}$ is a solution.

2nd Law: Angular momentum is conserved.

$$\therefore\ r^2\frac{d\phi}{dt} = C' = \text{constant}. \tag{4.2.12}$$

3rd Law: The sum of Kinetic plus potential energy is conserved.

$$\left(\frac{dr}{dt}\right)^2 + r^2\left(\frac{d\phi}{dt}\right)^2 - \frac{2GM}{r} = h' = \text{constant}. \tag{4.2.13}$$

For the comparison between GRT and Newtonian Mechanics results to be effective, we remember that in Newtonian Mechanics the proper time confounds itself with time "t" so that for the Newtonian metric we can write:

$$ds = cdt. \quad (4.2.14)$$

On a comparison we identify:

$$m = \frac{GM}{c^2}.$$

This result confirms the derivation done in section 4.1 (Formula (4.1.14)).

In Newtonian Mechanics, the solution for (4.2.12) and (4.2.13) is straightforward:

$$\frac{d^2u}{d\phi^2} + u = \frac{GM}{C'^2}, \quad (4.2.15)$$

where

$$u = \frac{1}{r}.$$

The Newtonian solution is a conic, given by:

$$u_N = \frac{1}{r} = \frac{GM}{C'^2}\left[1 + e\cos\phi\right], \quad (4.2.16)$$

where e stands for the eccentricity of the conic.

We now return to the relativistic equations which for our purpose of comparison can be written as:

$$\frac{dr}{ds} = \frac{dr}{d\phi}\frac{d\phi}{ds} = \frac{dr}{d\phi}\frac{C}{r^2} = -C\frac{du}{d\phi}, \quad (4.2.17)$$

and,

$$\left(\frac{du}{d\phi}\right)^2 + 2u\frac{du}{d\phi} - \frac{2m}{C^2}\frac{du}{d\phi} - 6mu^2\frac{du}{d\phi} = 0. \quad (4.2.18)$$

Differentiating the last equation in relation to ϕ, and applying equation (4.2.17), we obtain:

$$\frac{d^2u}{d\phi^2} + u = \frac{m}{C^2} + 3mu^2. \quad (4.2.19)$$

On comparison with (4.2.14) we identify:

$$\frac{m}{C^2} \equiv \frac{GM}{C'^2}.$$

The additional relativistic term,

$$3mu^2 = \frac{3GM}{c^2}\frac{1}{r^2},$$

turns out to be much smaller than the other terms, so it is a "perturbative" term.

We know then, that the relativistic orbit of the planet can only have a small deviation from an ellipse which is the Newtonian basic solution. In the perturbative term, we plug tentatively the Newtonian solution for the ellipse, u_N, so we write:

$$\frac{d^2u}{d\phi^2} + u \cong \frac{m}{C^2} + 3m(u_N)^2. \quad (4.2.20)$$

It is evident that the solution of the above equation must render a better approximation than the solution:

$$u \cong u_N,$$

for the equation (4.2.20).

We now substitute u_N by the value given in (4.2.16) and plug into (4.2.20); we make the approximation $e^2 << 1$, and find :

$$\frac{d^2u}{d\phi^2} + u = \frac{m}{C^2} + \frac{6m^3}{C^4} e\cos\phi, \tag{4.2.21}$$

where we also considered:

$$\frac{3m^2}{C^2} << 1.$$

The solution to the above equation is:

$$u = \frac{1}{r} = \frac{m}{C^2}\left[1 + e\cos\left(\phi - \frac{3m^2}{C^2}\phi\right)\right]. \tag{4.2.22}$$

On comparison between (4.2.22) and (4.2.16), we verify that there is an advance in the principal axis of the ellipse, which for a round turn ($\Delta\phi = 2\pi$) is given by:

$$\Delta\omega = \frac{6\pi m^2}{C^2}.$$

On the other hand, if we call "a" the semi major axis of the ellipse,

$$\frac{C^2}{m} = a(1 - e^2).$$

This is tantamount to:

$$\therefore \quad \boxed{\Delta\omega = \frac{6\pi GM}{c^2 a(1-e^2)}}. \tag{4.2.23}$$

For the planet Mercury,

$$\Delta w \simeq 0,104''.$$

In one century the calculated advance would be 43,0" as compared to the experimental value, from astronomical observations, 43,1".

We can not forget that the total advance in the perihelion of the planet Mercury is 5,600". The non-relativistic perturbations accounted by the interference of other planets' orbits accounts for 5,557" and only 43,41" are explained by GRT corrections, not justified in Newtonian Mechanics. This match was perhaps the most astonishing result of Albert Einstein's theory, and testifies in favor of GRT.

4.3. Propagation of Light near Gravitational Fields

In Special Relativity Theory the light trajectory is given by:

$$ds = 0.$$

Though this result is a bit obvious we shall demonstrate it for the sake of completeness, and because many beginning students do not find it explicit from some textbooks.

We recall Minkowski's metric:

$$ds^2 = -\left[dx^2 + dy^2 + dz^2\right] + c^2 dt^2. \tag{2.1.4}$$

If the speed of a particle is defined kinematically as:

$$V = \sqrt{dx^2 + dy^2 + dz^2}/dt,$$

and we apply this formula for a photon, we know that the result is $V = c$. Plugging back into (2.1.4) we find $ds = 0$. A result valid for photons and thus for light rays. When we go to GRT we are led to keep the equation for light rays as:

$$ds = 0. \tag{4.3.1}$$

For a photon in the neighborhood of a central mass M located at the origin of the coordinate system with a spherically symmetric distribution, we take Schwarzschild's metric, and the equations (4.2.19) and (4.2.11) can be kept along with (4.3.1). In our case the value of the constant C which appears in (4.2.11) is to be taken, obviously, as infinite:

$$C = \infty. \tag{4.3.2}$$

This being the case we are left from (4.2.19) with:

$$\frac{d^2u}{d\phi^2} + u = 3mu^2. \tag{4.3.3}$$

We take into account that the r.h.s. of (4.3.3) represents a small corrective or perturbative term, in the case of the Sun for example. In that case, $u = \frac{1}{r}$ would be calculated with r equal to the Sun's radius, and it is clear that the approximate solution of this equation would be $u \cong u_1$, where u_1 stands in the equation:

$$\frac{d^2u_1}{d\phi^2} + u_1 \cong 0.$$

Its solution is:

$$u_1 = \frac{\cos\phi}{R},$$

where R stands for a constant.

A better approximation may now be obtainable by solving the following equation:

$$\frac{d^2u}{d\phi^2} + u = 3m(u_1)^2 = \frac{3m\cos^2\phi}{R^2} \tag{4.3.4}$$

The reader should pay attention and focus himself that the above equation (4.3.4) comes in fact from (4.3.3) where in r.h.s. we have substituted u by its Newtonian solution.

The solution of (4.3.4) is

$$u = \frac{1}{R}\cos\phi + \frac{m}{R^2}\left(2-\cos^2\phi\right).$$

The equation which one finds for the deviation of light rays in the neighborhood of a central gravitational mass like the Sun, is obtainable if we compare the two farthest rays for which:

$$u = \frac{1}{r} = 0.$$

We have an equation of the second degree in $\cos\phi$:

$$(m/R)\cos^2\phi - \cos\phi - (2m/R) = 0.$$

Its roots are:

$$\cos\phi = \left[1 \pm \sqrt{1+\frac{8m^2}{R^2}}\right](2m/R)^{-1}.$$

In our approximation:

$$\frac{m^2}{R^2} << 1.$$

We now find:

$$\cos\phi \cong \left[1 \pm \left(1+\frac{4m^2}{R^2}\right)\right](2m/R)^{-1}.$$

The first solution is:

$$\cos\phi \cong (2m/R).$$

The second solution is:

$$\cos\phi' \cong \left(2+\frac{4m^2}{R^2}\right)(2m/R)^{-1} = \frac{R}{m} + \frac{2m}{R}.$$

The physically acceptable solution is the first one, because in this solution the deflection of light will be small accordingly to what one expects when we employ GRT instead of Special Relativity. This confirms that GRT offers small corrections to both S.R. and Newtonian Mechanics when we are not facing extremely powerful gravitational fields like for instance it would happen in the case of the gravitational collapse of an astrophysical object into a black-hole.

As in S.R. we would find $\cos\phi = 0$, we take now the solution:

$$\cos\phi << 1.$$

The final result is:

$$\phi \cong \pm\left(\tfrac{\pi}{2}+\tfrac{2m}{R}\right).$$

This angle is valid for the two farthest extremes of the light path, bringing a total deviation of:

$$\Delta\phi = \tfrac{4m}{R}. \tag{4.3.5}$$

The predicted theoretical deviation for the Sun, with a light ray passing close to its exterior, is thus:

$$\Delta\phi = 1.75''.$$

We may compare the above theoretical value with the experimental measurement for the Quasar 3C279 which is blackened by the Sun once in a year: the radio waves emitted by this object showed a deviation:

$$\Delta\phi = 1.73'' \pm 0.05''.$$

The predicted value is 97% accurate !!!

4.4. Spectral Displacement Due to Gravitational Cause

From a star of gravitational mass M and radius R_1 a certain electromagnetic radiation is emitted. Consider an observer located at a distance R_2 from the origin of the coordinate system which coincides with the center of the star, which we suppose homogeneous and spherical. We shall suppose that $R_2 >> R_1$. The Schwarzschild's metric is applicable:

$$ds^2 = \frac{-dr^2}{1-\frac{2m}{r}} - r^2 d\theta^2 - r^2 \sin^2\theta d\phi^2 + (1-\tfrac{2m}{r})dt^2. \tag{4.1.12}$$

Throughout this book we have not dealt with proper time τ up to now. It should not be identified with the variable "t" which stands for coordinate time. We define the former by:

$$d\tau = \tfrac{ds}{c}, \qquad \text{with} \qquad dr = d\theta = d\phi = 0.$$

Accordingly we shall have:

$$d\tau = \tfrac{ds}{c} = \sqrt{1-\tfrac{2m}{R}}\,dt,$$

(with $dr = d\theta = d\phi = 0$).

Let us suppose that at coordinate times t_1 and $t_1 + dt_1$, two successive wave crests are emitted from the surface of the star. The proper period of emission $d\tau_1$ is given by:

$$d\tau_1 = \sqrt{1-\tfrac{2m}{R_1}}\,dt_1. \tag{4.4.1}$$

If the observer, say, on the Earth's surface receives the impulses with a time difference dt_2, the proper period of reception $d\tau_2$ is given by:

$$d\tau_2 = \sqrt{1 - \tfrac{2m}{R_2}}dt_2. \qquad (4.4.2)$$

We are in face of a certain sort of "comoving" observers, which for the time being we define as observers static in relation to the source. We further consider that the period $dt_2 << 1$, and so, we commit a forgivable mistake if we suppose that:

$$dt_1 \cong dt_2. \qquad (4.4.3)$$

If λ stands for the emitted wavelength, and $(\lambda + \Delta\lambda)$ is the received corresponding value, we shall have:

$$\tfrac{(\lambda+\Delta\lambda)}{\lambda} = 1 + z = (1 - \tfrac{2m}{R_2})^{1/2}(1 - \tfrac{2m}{R_1})^{-1/2} \cong (1 - \tfrac{2m}{R_1})^{-1/2} \cong 1 + \tfrac{m}{R_1}. \qquad (4.4.4)$$

In the year 1965, Pound and Snyder made a precise measurement of a gravitational redshift in the Earth's gravitational field, by means of the Mössbauer effect. This experiment has been repeated with verification of formula (4.4.4) with 99% of confidence (see Pound and Rebka (1960)).

4.5. Other Verifications of the GRT

1st) Gravitomagnetism and galactic rotations

GRT has been verified exhaustively for about 100 years and no discrepancies have been found up to now. Gravitomagnetic phenomena were predicted for a long time until, Ignazio Ciufolini and others (see Appendix), reported recently that satellites carrying gyroscopes had verified closely the approximate solution (Lense-Thirring metric) given by the Austrian scientists Lense and Thirring at the beginning of 20th century. It turns out that in the same way as charged moving particles create a magnetic field, rotating masses create an additional "gravitomagnetic effect" (see Appendix). Likewise, Cooperstock and Tieu (2005) solved the galactic rotational problem, by means of GRT, finding that there is no use in imposing the existence of exotic "dark" matter, in order to explain the high tangential speeds of orbiting stars around, and far from, the center of the galaxy (see Appendix).

2nd) Hafele-Keating experiment:

In section 4.4 we mentioned the relation between proper time $d\tau_1$ and coordinate time dt for Schwarzschild's metric:

$$d\tau_1 = \sqrt{1 - \tfrac{2m}{R}}dt, \qquad (4.5.1)$$

in the presence of a gravitational field. On the other hand, in Special Relativity there is a second variation between proper time $d\tau_2$ and the coordinate time dt for a moving object with speed v :

$$d\tau_2 = \sqrt{1 - v^2/c^2}dt. \qquad (4.5.2)$$

Equation (4.5.1) tells us that clocks go slower in the presence of strong gravitational fields. Relation (4.5.2) tells us that clocks go faster when they move, relative to static clocks. By means of Hafele-Keating experiment it was possible to compare clocks readings inside airplanes with similar clocks on the Earth's surface. A team of the University of Maryland in the United States, made measurements between the years 1975 and 1976. Due to differences in the intensity of the gravitational fields on the Earth's surface, and the flying airplanes it was found an advancement in time measured by the clocks inside airplanes according to (4.5.2). However, this effect was partially camouflaged by the opposite effect due to the speed of the airplanes relative to static clocks on the Earth's surface. By employing slow motion airplanes, this opposite effect was reduced to less than 10% of the total retardation; thus formula (4.5.1) could be verified experimentally with 98% of precision relative to the theoretical value.

3rd) Shapiro experiment:

In the year 1965, the American scientist I. I. Shapiro proposed the following experiment: a radar wave emission should be directed to another planet so that the light ray would pass near the Sun. The planet would reflect the radar emission, the time lag would be measured and obviously the time taken would be larger than expected due to the passage near the Sun. In the year 1970 Shapiro made the measurement confirming Einstein's theory with 97% of accuracy.

4th) Hulse-Taylor observations:

As we mentioned in last Chapter, GRT's predictions of the existence of gravitational waves, got indirect confirmation by means of the observation of a Pulsar, with a possible companion; they would loose energy due to gravitational radiation according to the general relativistic calculation (Einstein's formula), not derived or shown in this book.

Note on gravitational radiation: it is my opinion that gravitational waves' direct detection, may be hindered by a misleading interpretation of the tensors formulae; there is a great difference between "tensor components", and "physical tensor components". The latter, are the ones that are experimentaly measured, and they may be affecting the order of magnitude of the sought slight relative displacement on the detectors. (Synge and Schild, 1969).

5th) Global Positioning System (GPS) — NAVSTAR :

By continuously emitting radio waves, at fixed frequencies, several orbiting satellites can active a GPS detector, (for instance, inside an airplane), which arrive with a Doppler shift, dependent on the relative speed of the source and the receiver; it also depends on the gravitational potential difference between the locations of the source and receiver. Both Special and General Relativity are involved; speeds and positions of the airplanes are then measured with extreme accuracy ($\pm$ 2 cm.s^{-1} ; 16 m). This system of navigation is called GPS-NAVSTAR, and works well.

Final Comments

Einstein has commented that it would suffice a single negative result breaking GRT's predictions, in order that this theory be discarded, despite all positives other confirmatory results.

Chapter 5

Complements of Tensor Calculus and General Relativity (Optional Study)

5.1. Orthogonal Transformations and Cartesian Tensors

Two coordinate systems S and S' relate themselves by means of an orthogonal transformation, if it can be expressed like:

$$x'^i = a_{ij}x^j + b^i, \qquad (5.1.1)$$

where a_{ij} and b^i are constants.

In a orthogonal transformation, the distance between two points, x_0^i and x^i is defined by its square:

$$s'^2 = (x'^i - x_0'^i)(x'^i - x_0'^i),$$

which is an invariant, i.e.:

$$s'^2 = s^2. \qquad (5.1.2)$$

We define:

$$\Delta x^i = x^i - x_0^i,$$

and we verify that (5.1.1) can be written like:

$$\Delta x'^i = a_{ij}\Delta x^j, \qquad (5.1.3)$$

or in matrix form:

$$\Delta x' = A\Delta x. \qquad (5.1.4)$$

If we call the transpose of $\Delta x'$, as $(\Delta x')^T$, we may write:

$$(\Delta x')^T(\Delta x') = (\Delta x)^T(\Delta x) = s^2. \qquad (5.1.5)$$

On taking the transpose of both sides of (5.1.4) we find:

$$(\Delta x')^T = (\Delta x)^T A^T. \tag{5.1.6}$$

We now combine relations (5.1.4), (5.1.5) and (5.1.6), finding:

$$(\Delta x)^T A^T A(\Delta x) = (\Delta x)^T (\Delta x). \tag{5.1.7}$$

From the above equation, we find:

$$A^T A = I, \tag{5.1.8}$$

where the symbol I stands for the identity matrix.

By taking determinants of both sides of the last equation, we find:

$$(\det A)^2 = 1,$$

because,

$$\det A^T = \det A.$$

We conclude that :

$$\det A = \pm 1. \tag{5.1.9}$$

From (5.1.8) it results also that:

$$A^T = A^{-1}, \tag{5.1.10}$$

where A^{-1} stands for the inverse of the matrix A.

The coefficients a_{ij} of the matrix A, in consequence, obey to relation:

$$a_{ji} a_{jk} = \delta_{ik}, \tag{5.1. 11}$$

due to relation (5.1.8). They also obey to the relation:

$$a_{ji} a_{ki} = \delta_{jk}, \tag{5.1. 12}$$

due to obvious relation:

$$AA^T = I,$$

which is obtained from (5.1.10) by multiplying both sides of the equation with the matrix A:

$$AA^T = AA^{-1} = I.$$

Relation (5.1.11) and (5.1.12) are the necessary conditions for the transformation to be orthogonal, and each one of them is a sufficient condition for this transformation.

Cartesian Tensors

The definition of tensors given on Chapter 1, can be simplified for orthogonal transformation by means of the relation:

$$\frac{\partial x'^i}{\partial x^j} = a_{ij}. \tag{5.1.13}$$

The above was obtained from (5.1.1). Notice that there is no distinction among covariant, mixed or contravariant components, because:

$$\frac{\partial x^i}{\partial x'^j} = \frac{\partial x'^i}{\partial x^j} = a_{ij}. \tag{5.1.14}$$

5.2. Parallel Transport in Riemann Space

We have been defining tensors and operations with tensors over all space. Nevertheless, in some occasions, it is convenient to deal with a vector $A^i(s)$ defined only on a curve $x^k = x^k(s)$, where "s" is a curve parameter, that we shall preferably associate with the arc length. Obviously there is no sense in speaking on its covariant derivative, $\frac{DA^i}{Dx^k}$ over all space, relative to a variable x^k. Nevertheless, we can talk of the covariant, or "absolute", derivative of the vector along the curve of parameter "s" which we shall denote by $\frac{DA^i}{Ds}$.

By definition,

$$A'^i(s) = \frac{\partial x'^i}{\partial x^j} A^j(s). \tag{5.2.1}$$

By the rules of differentiation, we have:

$$\frac{dA'^i}{ds} = \frac{\partial x'^i}{\partial x^j}\frac{dA^j(s)}{ds} + \frac{\partial^2 x'^i}{\partial x^k \partial x^j} A^j(s)\frac{dx^k}{ds}. \tag{5.2.2}$$

By analogy with the study already done, for the definition of covariant derivative, in Riemann space, it is clear that we may define along a given curve, the absolute derivative as follows:

$$\frac{DA^i}{Ds} \equiv \frac{dA^i}{ds} + \Gamma^i_{jk}\frac{dx^k}{ds}A^j. \tag{5.2.3}$$

Analogously, we could define covariant derivatives or absolute derivatives along a given curve, for any kind of tensor. For any kind of tensor, say T^{ij}_k we may define, by formal manipulation a relation between the absolute derivative along a curve and its covariant derivative over all space, even if it does not exist in practice:

$$\frac{DT^{ij}_k}{Ds} \equiv \frac{DT^{ij}_k}{Dx^m} \cdot \frac{dx^m}{ds}.$$

It must be remembered, however, that the covariant derivative along a curve may be the only one that we may calculate.

We define now the **parallel transport** of a tensor along a given curve, as the law which makes that its covariant derivative along the curve be zero. For instance, consider a contravariant vector A^i. The law of its parallel transport is given by:

$$dA^i = -\Gamma^i_{jk} A^j dx^k. \tag{5.2.4}$$

In order to understand the physical meaning of this law, consider parallel transport in a geodesic system of coordinates:

$$dA^i = 0.$$

Obviously its covariant derivative is also null:

$$\frac{DA^i}{Ds} = 0.$$

(in this coordinate system)

Thus, the vector does not change its size or direction. As the tensor properties are independent of the particular coordinate system to be employed, what we have said, is valid for the general case: the transported vector "does not change its size or direction". By means of parallel transport, we may compare two tensors located at different points of spacetime.

We may now interpret anew the equations of geodesics. As we saw in Chapter 3, it is given by:

$$\frac{d^2x^i}{ds^2} + \Gamma^i_{jk} \frac{dx^j}{ds} \frac{dx^k}{ds} = 0.$$

We rewrite the above as:

$$\frac{D}{Ds}\left(\frac{dx^i}{ds}\right) = 0. \tag{5.2.5}$$

Or:

$$\frac{D}{Ds} t^i = 0, \tag{5.2.6}$$

where $t^i = \frac{dx^i}{ds}$ is the vector tangent to the curve $x^i = x^i(s)$ by definition.

We are now in condition to say that the geodesics equation warrants that the tangent vector to geodesic curve is parallel-to-itself transported along the curve.

5.3. Fermi-Walker's Transport

When the observer does not move along a geodesic, the tangent vector $t^i = \frac{dx^i}{ds}$ will not coincide at another point of the curve with its paralleled transported vector. There is another transport law, the Fermi-Walker's, that assures that the Fermi-derivative:

$$\frac{D_F A^i}{Ds} \equiv \frac{DA^i}{Ds} - A_j \left[\frac{dx^i}{ds} \frac{D^2 x^j}{Ds^2} - \frac{dx^j}{ds} \frac{D^2 x^i}{Ds^2} \right], \tag{5.3.1}$$

is zero along the curve.

If the curve is a geodesic, $\frac{D_F}{Ds}$ coincides with $\frac{D}{Ds}$: Fermi-Walker's transport law for a geodesic coincides with parallel transport.

In order to show that the tangent vector to a curve transport itself along Fermi-Walker, the reader can check that, in (5.3.1) with:

$$A^i = \frac{dx^i}{ds},$$

we obtain:

$$\frac{D_F}{Ds}\left(\frac{dx^i}{ds}\right) = 0.$$

On the other hand, write $\frac{D^2x^n}{Ds^2} = 0$ in (5.3.1) in order to obtain:

$$\frac{D_F}{Ds}\left(\frac{dx^n}{ds}\right) = 0\,.$$

In other words, if a particle moves along a geodesic, the laws of parallel transport and Fermi-Walker's transport, coincide. We interpret (5.3.1) as follows: Fermi-Walker's transport law allows one to measure the variation of a vector A^i along a curve, departing from an initial point. It can be shown that the scalar product of vectors given by the expression $g_{ij}A^iA^j = A_jA^j$ is not altered by Fermi-Walker's transport, so that, lengths and vectors are preserved.

5.4. Lie Derivative

We now define a new type of tensor derivative called the Lie derivative. Consider an infinitesimal transformation of coordinates:

$$x'^i = x^i + \varepsilon a^i(x), \tag{5.4.1}$$

where ε is an infinitesimal parameter and $a^i(x)$ is a contravariant vector field which may be defined by:

$$a^i(x) = \left(\frac{\partial x'^i}{\partial \varepsilon}\right)_{\varepsilon=0} \tag{5.4.2}$$

Let now $T(x)$ be any tensor of any order, defined in each point of spacetime. Denote by P and Q two neighboring points with coordinates (x^i) and $(x^i + \varepsilon a^i)$ respectively. In point Q, the tensor field T may be calculated by two distinct modes:

1st mode: $T = T(x'^i)$. In this mode we have a value given by the definition of the field in each point including point Q.

2nd mode: $T' = T'(x'^i)$. In this mode we obtain $T'(x'^i)$ by means of $T(x^i)$ by application of definition of a tensor transformation from system of coordinates $x \to x'$, i.e., $P \to Q$.

We now define:

$$L_{a^i}T(x) = \lim_{\varepsilon\to 0}\left[\frac{T(x') - T'(x')}{\varepsilon}\right], \tag{5.4.3}$$

and call this, the Lie derivative of tensor T(x) in the direction of vector $a^i(x)$, at point Q.

Alternatively, because points P and Q coincide in the limit, we could have defined the same Lie derivative as above by means of:

$$L_{a^i}T(x) = \lim_{\varepsilon\to 0}\left[\frac{T(x)-T'(x)}{\varepsilon}\right]. \tag{5.4.4}$$

The two definitions are equivalent; the Lie derivative measure the variation of a tensor when the observer moves from one point to a neighboring point in a given direction, retaining its coordinate system.

Example 1: Calculate $L_a\phi(x)$, where ϕ is an invariant,

Consider that:

$$\phi(x) = \phi'(x'). \tag{5.4.5}$$

Let us apply the first definition: on point Q we have:

$$T(x') = \phi(x') = \phi(x+\varepsilon a) = \phi(x) + \varepsilon a^i \frac{\partial\phi}{\partial x^i},$$

and,

$$T'(x') = \phi'(x') = \phi(x)\ .$$

Then,

$$L_a\phi(x) = \lim_{\varepsilon\to 0}\left[\phi(x)+\varepsilon a^i\frac{\partial\phi}{\partial x^i} - \phi(x)\right]\varepsilon^{-1} = a^i\frac{\partial\phi}{\partial x^i}. \tag{5.4.6}$$

Example 2: Calculate $L_aV^i(x)$, i.e., obtain the Lie derivative of a contravariant vector.

Let us apply the second definition:

$$V'^j(x') = V'^j(x^i+\varepsilon a^i) = V'^j(x) + \varepsilon a^i\frac{\partial V'^j}{\partial x^i} \cong V'^j + \varepsilon a^i\frac{\partial V^j}{\partial x^i}, \tag{5.4.7}$$

where in the last term, we neglected second order infinitesimals ($\varepsilon^2 \cong 0$). From the definition of a contravariant transformation,

$$V'^j(x') = V^k(x)\frac{\partial x'^j}{\partial x^k}. \tag{5.4.8}$$

On the other hand,

$$\frac{\partial x'^j}{\partial x^k} = \frac{\partial}{\partial x^k}\left[x^j+\varepsilon a^j\right] = \delta^j_k + \varepsilon\frac{\partial a^j}{\partial x^k}. \tag{5.4.9}$$

From the last two relations, we obtain:

$$V'^j(x') = V^k \delta_k^j + \varepsilon V^k \frac{\partial a^j}{\partial x^k} = V^j + \varepsilon V^k \frac{\partial a^j}{\partial x^k}. \tag{5.4.10}$$

From (5.4.7) and (5.4.10), we find:

$$V'^j + \varepsilon a^i \frac{\partial V^j}{\partial x^i} = V^j + \varepsilon V^k \frac{\partial a^j}{\partial x^k} .$$

Then,

$$L_a V^i(x) = \lim_{\varepsilon \to 0} \left[\varepsilon \left(a^i \frac{\partial V^j}{\partial x^i} - V^k \frac{\partial a^j}{\partial x^k} \right) \right] \varepsilon^{-1} = a^i \frac{\partial V^j}{\partial x^i} - V^i \frac{\partial a^j}{\partial x^i}. \tag{5.4.11}$$

Analogously, we could find the Lie derivatives of any other kind of tensor.

Important result: in a Riemann space, the reader might check that an alternative but equivalent expression for relation (5.4.11) could be obtained by substituting partial derivatives by covariant derivatives; this conclusion is general and allows us to infer that the Lie derivative of a tensor, in Riemann space, is also a tensor.

Further results.

1st) The reader may check that:

$$L_a V_n = V_{n;i} a^i + V_i a^i_{;n}.$$

$$L_a g_{mn} = a_{m;n} + a_{n;m}, \tag{5.4.12}$$

where g_{mn} is the metric tensor component.

2nd) it obeys the usual rules for the derivative of sums and products.

3rd) it commutes with the contraction.

4th) it commutes with partial derivatives.

5.5. Isometries

The study of the symmetries in Riemann space, begins with the calculation of the variation of the metric tensor in given directions of space.

Definition: Isometric mapping is an infinitesimal coordinate transformation under which the Lie derivative of the metric tensor becomes null:

$$L_a g_{ij}(x) = 0. \tag{5.5.1}$$

On remembering relation (5.4.12), we find then:

$$a_{i;j} + a_{j;i} = 0. \tag{5.5.2}$$

This is the famous Killing equation, whose solutions, if they exist, allow the isometric mapping. The solutions $a_i(x)$ are the Killing vectors, and if they exist, we say that the space has intrinsic symmetries.

Example: Find the intrinsic symmetries of the Euclidean plane, i.e., solve Killing's equations for the Euclidean plane.

Solution:

In Cartesian coordinates,

$$ds^2 = dx^2 + dy^2$$

$$\therefore g_{AB} = \delta_{AB} \quad \text{where} \quad A, B = 1, 2.$$

Killing equation reduces to:

$$\frac{\partial a^1}{\partial x} = \frac{\partial a^2}{\partial y} = 0. \tag{5.5.3}$$

and,

$$\frac{\partial a^1}{\partial y} + \frac{\partial a^2}{\partial x} = 0. \tag{5.5.4}$$

By integration of the first equation (5.5.3) we find:

$$a^1 = Y(y).$$

and,

$$a^2 = X(x).$$

On taking to the second equation (5.5.4) the above results become:

$$\frac{dY(y)}{dy} + \frac{dX(x)}{dx} = 0.$$

This last equation can be solved by the method of separation of variables:

$$\frac{dY(y)}{dy} = -\phi \quad = \text{ constant.}$$

$$\frac{dX(x)}{dx} = +\phi \quad = \text{ constant.}$$

With the following solutions:

$$Y = -\phi y + x_0. \tag{5.5.5}$$

$$X = +\phi x + y_0,$$

where (x_0, y_0) is a pair of integration constants.

Geometrical interpretation of the above result:

We have three degrees of freedom in the solution, which describe the infinitesimal group of motions of the bidimensional Euclidean space; these are the translations x_0 and y_0 along the axes OX and OY, and rotations around the origin $x = y = 0$ according to an angle arc sin ϕ.

5.6. Stationary and Static Fields

A gravitational field is said to be stationary if it admits a Killing temporal vector, i.e., $a_i a^i > 0$. From the Killing equation:

$$a_{i;j} + a_{j;i} = 0,$$

where,

$$a_i = g_{ij} a^j.$$

We shall now investigate the above definition. We build a coordinate system such that only the coordinate x^0 varies along the trajectories of a^i keeping constant the space coordinates (x^1, x^2, x^3). In this system the trajectories of a^i are the self axis x^0, and $a^1 = a^2 = a^3 = 0$. On choosing a unit Killing vector,

$$a^i(x) \equiv (1,0,0,0),$$

we shall find Killing's equation reduced to:

$$a^i \frac{\partial g_{jk}}{\partial x^i} = a^0 \frac{\partial g_{jk}}{\partial x^0} = 0, \text{ or,}$$

$$\boxed{\frac{\partial g_{jk}}{\partial x^0} = 0}$$

In the new coordinate system all the components of the metric tensor are time independent. We say that the coordinate system is adapted to the stationary character of the metric, i.e., it is adapted to the Killing vector field. Kerr's metric which describes a rotating body is an example of stationary metric.

Static Field

A given gravitation field is called static, if there exists a coordinate system adapted to the stationary character of the metric for which

$$g_{0\nu} = 0 \quad (\nu = 1,2,3) \text{ and}$$

$$\frac{\partial g_{ij}}{\partial x^0} = 0.$$

The geometric interpretation is that the trajectory of the Killing vector fields $a^i(x)$ are, for a static field, orthogonal to a family of hypersurfaces.

Mathematical Note.

Given an equation of the type:

$$F(x^1, \ldots, x^N) = 0,$$

we say that it determines an hypersurface in N-dimensional space. In fact, by the rules of multi-variable Calculus:

$$dF = 0 \quad \therefore \quad \frac{\partial F}{\partial x^i} dx^i = 0.$$

In other words, for each displacement on the surface, the gradient vector $\frac{\partial F}{\partial x^i}$ which is covariant, is orthogonal to the hypersurface. In that way, the Killing vector $a^i(x)$ will be orthogonal to a given family of hypersurfaces $\sigma(x) = 0$ if it can be expressed by a relation of the type:

$$a_i(x) = \chi(x) \frac{\partial \sigma(x)}{\partial x^i} \, ,$$

where $\sigma(x)$ and $\chi(x)$ are scalar functions. Schwarzschild's metric is static according with this definition.

Part III

Introduction to Relativistic Cosmology

Chapter 6

Digression into Philosophical and Mathematical Matters (Optional Study)

6.1. Newtonian Concepts of Space and Time

Newton professed a philosophical conception over space and time, originated from the theories of Henry More (Koyré, 1962). Notwithstanding Newton having looked how to justify his space conception in the context of Physics, with the critical analysis of the notion of inertial force, he employed Henry More's philosophical system as a supplemental argument which seems made in order to reinforce his scientific conception. To Henry More, space and time should be absolute and infinite because they were manifestation of a "divine entity". This does not seem to us a serious argument because it mixes the quantitative infinity with the ontological infinity and because it materializes time and space along with the "divinity". More than that, as we shall show later, the philosophical analysis of gravitational and inertial phenomena shows that spacetime is a form connected with matter. This analysis, which encompasses Mach's Principle, was the departure point from which Einstein developed his covariant theory of Gravitation, which is known as the GRT (General Relativity Theory).

A basic principle of Newtonian dynamics, and also of the Special Theory of Relativity, is the relativity of the inertial systems of references. This means: there are no privileged reference systems, or, in other words, each one is equivalent to any other. We can say that the dynamical laws in those theories are invariant under inertial reference systems. This principle does not apply to accelerated reference systems. Therefore, Newton's second law, which says that force is the temporal derivative of linear momentum, is not anymore universally valid. This means that for coordinate transformations involving accelerated observers, there is a surge of other kinds of force, the so called inertial forces. The universal character of dynamical laws is now broken. More than that, because Newton's second law represents the principle of efficient causality ("give me the time derivative of linear momentum and I will show you the force"), now there is a problem as to the validity of this principle which was pointed by Einstein (1923): why should exist forces responsible for mere coordinate transformations and which do not respond for any known interaction?

Let us examine what happens with the rotation motion of a particle. Consider two observers Σ and Σ' one in rotation relative to the other. If the relativity principle would be valid, each one of both observers could tell to the other one: "I am at rest, and you are in rotational motion". This is not verifiable in Newtonian physics because one of those observers faces acceleration, while the other one does not. Consequently, Newton, in order to save the relativity principle, postulated the existence of an absolute observer so that he could distinguish Σ from Σ'. The presence of an acceleration in Σ' and not in Σ, is due to the existence of an absolute space, relative to which a given referential can either be accelerated or not. But as Einstein (1923) observed, this does not put the principle of causality in a safe ground. We conclude that, within the Newtonian theory of absolute space, the principle of causality becomes mutilated. The reader should remember that Einstein substituted, in Special Relativity, absolute time, and absolute space, for an absolute spacetime, represented by an invariant line element (Minkowski's). In General Relativity Theory, this "absolutism", was reduced to a local concept, in a Riemannian space.

In the next section we shall follow arguments against a theory of cosmology made up from Newtonian gravitational theory.

6.2. Objections to Newton's Cosmology

The basis for Newtonian Cosmology is Poisson's equation for the gravitational field. For readers not familiar with this equation we refer to basic university physics texts, where, in electrostatics, there is an analogous equation:

$$\nabla^2 U = 4\pi G\rho. \tag{6.2.1}$$

The above equation for the Universe, presupposes that matter is continuously distributed with mass density ρ, while G stands for Newton's gravitational constant and U is the gravitational potential. We justify the matter hypothesis above, because the mass distribution is extended over cosmic distances.

Einstein showed that this equation does not substitute perfectly the action-at-a-distance gravitational law because of the boundary conditions for the solution of the equation. In other words, we have to impose boundary conditions for the gravitational potential U at infinity. The solution of Poisson's equation requires that, when $r \longrightarrow \infty$, we should have $\lim_{r\longrightarrow\infty} U = U_\infty =$ fixed value. We conclude that $\lim_{r\longrightarrow\infty} \rho = 0$.

Nevertheless, we know that $U \sim r^{-1}$ (Halliday, Resnick and Walker, 2005). So $\rho \sim r^{-3}$. This means that ρ goes to zero more rapidly than U, for large values of r. This behavior led Einstein to say that Newtonian cosmology models possesses infinite mass, but in some sense it is finite because the gravitational potential goes to zero more slowly than the density!!! Let us now examine why Poisson's equation was not acceptable to Einstein in the cosmological domain.

First: the inhomogeneity due to $\rho \sim r^{-3}$ is incompatible with astronomical observations.

Second: it includes the notion of "center of the Universe" which is meaningless.

Third: Einstein shows that $\rho \sim r^{-3}$ is against statistical physics theory. The boundary condition imposes - as we saw – to Poisson's equation, that $\lim_{r\to\infty} U = U_\infty =$ fixed value. Now, by statistical physics, each astronomical object could acquire kinetic energy enough to overcome this value and then disappear towards infinity. Thus we would not find $\rho \sim r^{-3}$. We see therefore, that the static character of this Universe which is represented by Poisson's equation, is inviable.

Fourth: We could find an alternative solution $\rho =$ constant, but this can not be sustained by Poisson's equation because the gravitational potential U would become undetermined. This objection was presented by Seeliger in the year 1895, as cited in North (1965).

Fifth: In order to make superseded, the Seeliger's objection, it could be suggested that there is a universal constant λ , in a modified Poisson's equation, namely:

$$\nabla^2 U - \lambda U = 4\pi G\rho. \tag{6.2.2}$$

A possible solution to this equation would be:

$$U = -\tfrac{4\pi G}{\lambda}\rho, \tag{6.2.3}$$

with $\rho =$ constant.

However this modification introduces a new constant λ by means of an ad-hoc hypothesis. This runs against the theoretical framework.

Sixth: As Einstein observed, the radiating energy is originated by mass and it gets lost into infinity, which is also incompatible with the hypothesis $\rho \sim r^{-3}$.

Seventh: The functions $U \sim r^{-1}$, and $\rho \sim r^{-3}$, carry a nonsense: at infinity we would have a constant value U_∞ and the absence of mass because ρ goes to zero faster than U goes to U_∞. Einstein interpreted this hypothesis as being a universe with infinite mass, contained in a finite volume. Obviously, the mathematics is inconsistent: a universe in which $\lim_{r\to\infty} U = U_\infty =$ constant, with $\rho_\infty = 0$, while $U \sim r^{-1}$ and $\rho \sim r^{-3}$, gives us the idea that there is an exterior space outside the universe, because in physical theory we can not speak rigorously about infinities. When we say that $\lim_{r\to\infty} \rho = 0$, we simply represent mathematically the idea that for large distances r, the density goes to zero. Therefore such infinite limit brings about the concept of exterior space around the Universe, whose intrinsic antinomy was pointed by Nicolaus Cusa (1954), Copernicus, Descartes (Koyré, 1962) and others.

The idea of exterior space outside the Universe implies the concept of a Universe limited by a surface separating matter from vacuum, which is a senseless proposition. In fact, space as such, is a set of all possible positions of bodies relative to each others, i.e., a common form of phenomena (Weyl, 1950). As Saint Augustine (1958) remarked, the place of the universe is in itself, and there is no sense in talking about the Cosmos' exterior.

6.3. Mach's Principle

We saw, in the last two sections, the objections that can be raised against Newton's conception of infinite and absolute space, as well as the contradictory character of a cosmology based in his gravitational theory. Ernst Mach (1912), at the end of the nineteenth century, saw the difficulty involved in the justification given by Newton to the existence of an absolute reference system independent of the matter field, based on the consideration of local inertial forces. Logically, Mach refuted Newton's conclusion, on grounds that a reference system can only have physical significance, if it is tied to a matter field. He concluded, therefore, that local inertial accelerations should be referred to an inertial cosmic frame, defined by the distant masses of the Universe.

Einstein (1923), considering Mach's criticism, and, taking the causality principle in the definition of forces into account, inferred that local inertial forces are determined by a gravitational interaction between the local system and the distribution of cosmic masses. This is the famous Mach Principle, as coined by Einstein. So, Mach Principle encompasses two fundamental elements, as cited by Brans and Dicke (1961) and Gomide (1973):

1) Local accelerations are related to an inertial cosmic reference system determined by the distribution of the (distant) masses of the Universe.

2) These accelerations, which are responsible for inertial forces, are the result of a gravitational interaction between the local system and the (distant) cosmic masses. Inertial effects are connected with the gravitational interaction.

Mach Principle as formulated by Einstein, requires that the content of the causality principle for the existence of forces can not be justified by a mere appeal to an absolute reference frame, because these forces have their *raison d' être* in the reality of a physical interaction. The explanation given by Newton, as we saw earlier, hurts a fundamental principle of ontology. Einstein's philosophical view is much more penetrating than that of his XVIIth century colleague. Robert Dicke (1967) observes that Einstein succeeded better, when he based his GRT in dense philosophical analyses.

Mach Principle implies that a cosmic reference frame is determined by the distribution of masses in the Universe. Now, relativistic physics shows us that the material processes occur in a four dimensional continuum. By consequence, time must be associated to any frame in the universe. This fundamental notion may induce a more penetrating view of the problem caused by the existence of inertial forces in local reference frames, and to the better understanding of the difference between of the proper time and coordinate time, as referred to Mach Principle.

Given one local reference frame not subject to accelerations we could employ cylindrical coordinates in order to define the usual metric of Special Relativity. Let it be:

$$d\tilde{s}^2 = -\left[d\tilde{r}^2 + \tilde{r}^2 d\tilde{\phi}^2 + d\tilde{z}^2\right] + c^2 dt^2. \quad (6.3.1)$$

Consider now a rotating frame, such that:

$$\begin{aligned} z &= \tilde{z} \\ r &= \tilde{r} \\ t &= \tilde{t} \\ \phi &= \tilde{\phi} - \omega t \quad . \end{aligned}$$

Let ω be the angular velocity, so that the new metric is given by:

$$ds^2 = -\left[dr^2 + r^2 d\phi^2 + dz^2\right] + 2r^2 \omega d\phi dt + (c^2 - \omega^2 r^2)dt^2. \quad (6.3.2)$$

Consider now two observers at rest in both reference frames, respectively; we shall have:

$$d\tilde{s}^2 = c^2 dt^2. \quad (6.3.3)$$

$$ds^2 = (c^2 - \omega^2 r^2)dt^2. \quad (6.3.4)$$

We see that the proper times in both frames are different from one another, and related by:

$$ds^2 = (c^2 - \omega^2 r^2)(d\tilde{s}^2/c^2). \quad (6.3.4.a)$$

We may infer that the problem which left Newton perplex, i.e., the impossibility of application of the relativity principle from inertial to rotating frames is originated by a property ignored at the time, namely that time is relativistic and not absolute. The duality between proper time and coordinate time is underlined in the grounds of rotating frames and it can be perceived by the privileged character of the frame under centrifugal acceleration; its consequence is the existence of a metric component $g_{00} \neq 1$, and given by:

$$g_{00} = (c^2 - \omega^2 r^2)/c^2, \quad (6.3.5)$$

which entails a local inertial effect. The accelerating reference frame possesses a proper time different than that of the non accelerating system. Now, Mach Principle, as seen from the relativistic point of view, must also mean the determination by the cosmic masses, of a common universal time with the local observer (Gomide,Berman, Garcia, 1986). Once that the time measurement is not independent of matter, the cosmic mass should be in a global motion (like, for instance, expansion) capable of defining such time. It must be observed that a static cosmic distribution of masses could not define a "time": for a "time" in a static universe would not be different than the Newtonian absolute time independent of matter, i.e., it could not be defined because, as we pointed above, time is tied to motion. What we saw, namely:

$$ds^2 = g_{00}(r)d\tilde{s}^2, \tag{6.3.6}$$

it is there to signify an interaction of the local rotating system with the cosmic system in motion, via the liaison between both proper times. Consequently the metric component $g_{00}(r)$ which represents an inertial effect is also significant of an interaction of the local accelerated system with the distribution of masses in the Universe. From this original visualization of Mach Principle, according to the relativity of time, we must conclude that for the principle to be realized, the cosmic masses must be in motion, unless we fall into the untenable idea of an absolute Newtonian cosmic time. The interaction property inherent to $g_{00}(r)$ attaches a special significance to the gravitational interaction potential, when we think of the equivalence principle (see section 6.5).

It can be shown, as we shall also point out in later Chapters, that a mathematical formulation of Mach's Principle may be assumed to be the statement, (and its mathematical consequences) that the total energy of the Universe is zero-valued. (Berman, 2006; 2006 a).

D'Inverno summarizes Machian ideas with three statements (D'Inverno, 1993):

I - the matter distribution determines geometry.
II - "no matter" means "no geometry".
III - in an empty Universe, a body would have no inertia.

6.4. The Potential for Gravitation

From Mach Principle, which denies the existence of Newtonian absolute space, and them postulates the equivalence of inertia and gravitation in the cosmic domain, it is only a question of mere necessity that a new formulation be made for the gravitational theory. We have seen that Newton's scalar potential presupposes the idea of an absolute space dissociated from matter, and without agreeing with any equivalence between inertia and gravitation. Dicke (1964) studied some possibilities for a possible gravitational potential that would obey Mach Principle, while showing that the best candidate is a tensorial potential associated with the metric tensor. Let us follow the line of thought established by Dicke:

First option: a scalar potential in a quadridimensional spacetime.

In this case a quadriforce acting on a particle due to a scalar field would be given by:

$$F_\alpha = -\phi_{,\alpha}. \tag{6.4.1}$$

Newton's second law should be:

$$F_\alpha = \frac{dp_\alpha}{d\tau}, \tag{6.4.2}$$

where p_α is the linear quadrimomentum.

By hypothesis, the scalar potential is not time-dependent so that:

$$\phi_{,0} \equiv \frac{\partial \phi}{\partial x^0} = 0. \tag{6.4.3}$$

Now, this means that the energy of the particle, i.e., the time component of the quadrimomentum, is:

$$p_0 = m(1-\beta^2)^{-1/2}c^2 = \text{constant}. \tag{6.4.4}$$

Now, the existence of a force acting on the particle means that the kinetic energy has to vary with time, which is contradictory to p_0 being a constant.

More than that, in order for p_0 to be constant it would be necessary that its rest mass should vary with time which contradicts the concept of the rest mass of a particle in Special Relativity.

Second option: Vector potential in quadri-dimensional spacetime.

This would be a potential in an analogy with that of the electromagnetic field. Such a vector potential would determine a privileged direction in space because the rotation part of the field carries field lines of a non isotropical paradigm. (The reader can easily think that there is one privileged direction in spacetime: that of the vector potential). Now, the local inertial effects posses an isotropic character. [Weinberg (1972) discusses nevertheless, the study of anisotropy of inertia].

Third option: Tensor potential.

In Classical Mechanics the equations of motion can be derived from Hamilton's Principle (Goldstein, 1980):

$$\delta \int [K-V]\, dt = \delta \int L dt = 0. \tag{6.4.5}$$

where K, V and L are the kinetic energy, potential energy and the Lagrangian of the problem. We would be tempted to write for the motion of a particle in the gravitational field an equation of the type:

$$\delta \int [I+G]\, d\tau = 0, \tag{6.4.6}$$

where $L = I + G$. (We separated the Lagrangian into two parts: the inertial I and the gravitational G). However, from Mach Principle, due to the equivalence of inertia and gravitation, we may convert I into G and vice-versa. So we would write:

$$\delta \int G d\tau = 0 \tag{6.4.7.a}$$

$$\delta \int I d\tau = 0 \quad . \tag{6.4.7.b}$$

Then we must look for a Lagrangian quadratic in velocity which represents the kinetic part *K*of the motion, but whose terms are multiplied by terms representative of the potential for gravitation or the inertia potential.

So, we must drop the relation $L = K - V$ with preference to a relation of the type $L = KV$. This product evidences the connection of the inertial effects of kinetic origin, with gravitation or inertia. Gravitation and inertia appear in a mixed form. Then this Lagrangian must assume the form:

$$L = g_{\alpha\beta}u^{\alpha}u^{\beta}. \tag{6.4.8}$$

This means that the gravitational or inertial potential is a second order tensor $g_{\alpha\beta}$. The most simple, (but representative of Mach Principle), variational principle would be:

$$\delta \int g_{\alpha\beta}u^{\alpha}u^{\beta}d\tau = 0. \tag{6.4.9}$$

Notice that we use proper time as the integration variable. This principle is formally equivalent to the equation of geodesics in Riemannian space:

$$\delta \int ds = 0, \tag{6.4.10}$$

from which, as it is well known, we derive the metric form of the geodesics,

$$ds^2 = g_{\mu\nu}dx^{\mu}dx^{\nu}, \tag{6.4.11}$$

which can also be written:

$$g_{\mu\nu}\frac{dx^{\mu}}{ds}\frac{dx^{\nu}}{ds} = 1. \tag{6.4.11.a}$$

We remember that the quadrivelocities are defined by:

$$u^{\alpha} = \frac{dx^{\alpha}}{d\tau}.$$

Mach Principle, therefore, suggests that the tensor potential for gravitation is equivalent to the Riemann metric tensor. So, gravitation and inertia determine the geometry of spacetime.

6.5. Equivalence Principle

Another principle which connects gravitation, inertia, and spacetime geometry, is the equivalence principle, which has an experimental basis. Let us call $\vec{g}$ – the acceleration of gravity; and m$_i$ the inertial mass of a particle, i.e., the mass that appears in Newton's second law. If m_g stands for the gravitational mass, which is the mass that appears in the acceleration of gravity or with Newton's Law of Gravitation, we may write:

$$m_g\vec{g} = m_i\vec{a}, \tag{6.5.1}$$

where $\vec{a}$ stands for the acceleration of the particle subject to the gravitational field. From the above we write:

$$\vec{a} = (m_g/m_i)\vec{g} \,. \tag{6.5.2}$$

It would seem that the acceleration $\vec{a}$ would be different for the different bodies with different values of m_g/m_i. Galileo in the sixteenth century verified, with the precision limits available at his time, that the acceleration $\vec{a}$ for bodies in free fall independs on the masses of those bodies. Newton, in the next century, tried pendulums of the same length, same mass and different materials, and did not find any difference in their respective periods. Friedrich Wilhelm Bessel, one century later, confirmed Newton's results with a better precision. In the last century, Roland von Eötvös (Eötvös et al, 1922) verified that the ratio m_g/m_i independs on the nature of those bodies, with a precision of 5 in 10^9. The experiment was repeated by Dicke (1964) with the precision 1 in 10^{11}, with the same result: the mentioned ratio independs on the nature of the bodies. Though all these experiments only prove that gravitational and inertial masses are proportional to each other, without loss of generality, we can set the proportionality as an equality, and write:

$$m_g/m_i = 1. \tag{6.5.3}$$

We can conclude that the path followed by a body subject to inertia and gravitation, independs on the formal atomic structure of the body. In this regard, inertia and gravitation are only related with the trajectory of a body in spacetime. Now, spacetime is the common ground for relationships among physical phenomena (Weyl, 1950), which means that inertia and gravitation are only connected with physical geometry. Saying that otherwise, the geometry of spacetime is the common form of the matter in the interaction of inertia with gravitation, which is, the universal property of matter; their properties can be expressed in terms of the geometry. To say that the path of a body is solely determined by gravitation and inertia, is the same thing as to say that this path is solely determined by the geometry. Which curves are those which only depend on spacetime? – These are the geodesics. In Euclidean geometry the geodesics are straight lines. But a body like a planet subject to gravitation and inertia, does not move in a straight line. This means that the geodesics must be a family of curves in non-Euclidean geometry. The obvious choice is Riemannian spacetime, whose geodesics are given by:

$$\delta \int ds = \delta \int \sqrt{g_{\alpha\beta}u^{\alpha}u^{\beta}}ds = 0. \tag{6.5.4}$$

This results in:

$$\frac{d}{ds}(g_{\alpha\beta}u^{\beta}) - \frac{1}{2}g_{\beta\delta,\alpha}u^{\beta}u^{\delta} = 0,$$

while

$$ds^2 = g_{\alpha\beta}dx^{\alpha}dx^{\beta}.$$

We may write, taking into account that $m_g = m_i$,

$$m_i \frac{du_\alpha}{ds} = \frac{1}{2} m_g g_{\beta\delta,\alpha} u^\beta u^\delta. \tag{6.5.5}$$

In this last form, we have a generalized Newton's second law where the inertial force is equivalent to a force arising from a gradient of the tensorial potential $g_{\alpha\beta}$. Now, the metric coefficient of the geometry has the physical meaning of a tensor potential. We have found the law of free fall in a uniform gravitation field, from the equations of geodesics.

D'Inverno (1993) states that the above implies also in five statements :

I - the motion of a small body in a gravitational field independs on its mass and composition.
II - matter is acted by gravitational fields, but is also a source of such fields.
III - the gravitational field is coupled to everything else because it contains energy.
IV - the concept of inertial frame is localized.
V - a rest frame immersed in a gravitational field, is equivalent to a reference frame linearly accelerated in Special Relativity.

6.6. The Principles of (General) Relativity, Covariance, and Simplicity

As Mach Principle denies the existence of privileged systems of reference, all observers must be equivalent, and not only the inertial ones. There should be a *principle of general relativity* for arbitrary transformations of coordinates. We are led to choosing the line element of Riemann geometry, namely:

$$ds^2 = g_{\mu\nu} dx^\mu dx^\nu,$$

because this is an invariant quantity for arbitrary coordinates.

We have seen earlier, that the metric tensor has to do with accelerations. This results from Mach and Equivalence Principles. These two principles make us consider the metric tensor of Riemann geometry as applicable to Physics in order to represent the gravitational field or the accelerations field. The coefficients of the metric tensor assume physical properties. This shows that Newtonian theory, which considers physical spacetime as an empty and absolute spacetime which has nothing to do with the matter fields, should be wrong. As in the Greek Philosophy of Aristotle and Plato space and time are seen as a category of matter.

The **principle of covariance**, for physical laws, may be thought as devoid of physical contents, for one could take, a given relation, valid only in a particular system, and write it in tensor form, but if the physics involved would not allow for a correct translation into a covariant phenomenon, the tensor equation would be devoid of validity. Nevertheless, correct physical laws must be written in tensor notation, which is the best way to know how to express it in other coordinate systems, with equal validity.

We have seen that, in Special Relativity, the energy momentum tensor obeys a conservation law:

$$T^{\mu\nu}_{,\nu} = 0. \tag{6.6.1}$$

In General Relativity, the obvious generalization would be:

$$T^{\mu\nu}_{;\nu} = 0. \tag{6.6.2}$$

However, there remain other possibilities, involving the curvature tensor, which is null in Special Relativity; for instance (D'Inverno, 1993):

$$T^{\alpha\beta}_{;\beta} + g^{\beta\varepsilon} R^{\alpha}_{\beta\gamma\delta} T^{\gamma\delta}_{;\varepsilon} = 0. \tag{6.6.3}$$

We point out that simplicity requirements imply in "minimal coupling", so that we prefer (6.6.2) as the correct generalization of (6.6.1): no terms including the curvature tensor should, then, be added in the generalization.

For consistency, we must impose a correspondence law, contained in the statement:

"General Relativity reduces to Special Relativity in the absence of gravitation; it reduces to Newtonian gravitation theory in the case of weak gravitational field and low speeds; and to Newtonian mechanics, in the absence of gravitation, and for low speeds".

By the same token, we do not admit, in General Relativity, variations, either in time or in space, of the constants in the theory:

$$\frac{\partial G}{\partial x^\alpha} = \frac{\partial m_0}{\partial x^\alpha} = \frac{\partial c}{\partial x^\alpha} = \frac{\partial \varepsilon_0}{\partial x^\alpha} = \frac{\partial \mu_0}{\partial x^\alpha} = 0. \qquad (\alpha = 0, 1, 2, 3)$$

In the above, $G, m_0, c, \varepsilon_0, \mu_0$ are respectively Newton's gravitational constant, rest mass, speed of light, permittivities of electric and magnetic origin.

Of course, removing the first term requirement makes for Brans-Dicke theory; the second one, for Kaluza-Klein-Wesson penta-dimensional theory; the other three, for Albrecht-Magueijo's. (Brans and Dicke, 1961; Wesson, 1999, 2006; Albrecht and Magueijo, 1998).

The requirement of simplicity, does not mean that we should ignore experimental observations; the cosmological constant, was dismissed from the theory some time ago, on this requirement: however, cosmological observations asked recently for its reinstatement. A variable G or Λ, may be necessary from physical requirements.

Riemannian geometry might eventually have to be replaced by a new geometry which included torsion, besides curvature (Berman and Marinho, 1996). The simplicity requirement could only be used to keep the first geometry, if there were no physical evidence to the contrary; again, the simplicity requirement could be used, for instance, to choose between the tridimensional possibilities for the physical world, and express that it is flat; or to decide between isotropic or anisotropic models of the Universe, in favor of the former; or between homogeneous or inhomogeneous models of the Universe, choosing the former, etc.

6.7. Principles of "Cosmic Time" and "Correspondence"

Consider the metric line-element:

$$ds^2 = g_{\mu\nu} dx^\mu dx^\nu \tag{6.7.1}$$

If the observer is at rest,

$$dx^i = 0 \qquad (i = 1, 2, 3) \quad ,$$

while,

$$dx^0 = dt \tag{6.7.2}$$

This last equality defines a proper time; we called cosmic time, in Cosmology.

From the geodesics' equations, we shall have:

$$\frac{d^2x^i}{ds^2} + \Gamma^i_{\alpha\beta} \frac{dx^\alpha}{ds} \frac{dx^\beta}{ds} = \Gamma^i_{00}. \tag{6.7.3}$$

We then find:

$$g^{ij} \frac{\partial g_{i0}}{\partial t} = 0. \tag{6.7.4}$$

This defines a Gaussian coordinate system, which in general implies that:

$$\frac{\partial g_{i0}}{\partial t} = 0. \tag{6.7.5}$$

We must now reset our clocks, so that, the above condition is universal (valid for all the particles in the Universe), and then our metric will assume the form:

$$ds^2 = dt^2 - g_{ij} \left(\vec{x}, t \right) dx^i dx^j. \tag{6.7.6}$$

If we further impose that, in the origin of time, we have:

$$g_{i0}(t = 0) = 0, \tag{6.7.7}$$

then by (6.7.5), we shall have:

$$g_{i0}(t) = 0. \tag{6.7.8}$$

The above defines a Gaussian normal coordinate system.

For a comoving observer, in a freely falling perfect fluid, the quadrivelocity u^μ will obey:

$$u^i = 0, \tag{6.7.9}$$

while, if we normalize the quadrivelocity, we find, from the condition:

$$g_{\mu\nu} u^\mu u^\nu = 1, \tag{6.7.10}$$

that,

$$g_{00}u^0 = 1. \tag{6.7.11}$$

Though Gomide and Berman (1988) have discussed the case $g_{00} = g_{00}(t) \neq 1$, we usually impose:

$$g_{00} = u^0 = 1. \tag{6.7.12}$$

The purpose we have in mind, is to define a Machian metric; Gaussian coordinate systems, in fact, imply that, with $g_{0i} = 0$, there are no rotations in the metric (we refer to formula (6.3.3) to the contrary), and in each point we may define a locally inertial reference system.

Gaussian normal coordinates were called "synchronous"; in an arbitrary spacetime, when we pick a spacelike hypersurface S_0, and we eject geodesic lines orthogonal to it, with constant coordinates x^1, x^2 and x^3, while $x^0 \equiv t + t_0$, where $t_0 = 0$ on S_0, then t is the proper time, whose origin is $t = 0$ on S_0 (see MTW, 1973).

In the above treatment, cosmic time is "absolute", so that the measure of the age of the Universe, according to this "time", is not subject to a relative nature.

As it was remarked in Chapter 3, a GRT formula involving covariant derivatives, reduces, in its Special Relativistic equivalent, to partial derivatives. The converse, however, may not be true. Also, to each point in a Riemannian space, there corresponds a "tangent" plane, where Special Relativity laws, are locally valid. This is not a global property. Geodesic coordinates, are meant to reproduce a nearly flat-space frame around a given point. If a tensor equation, is shown to be valid in such coordinates, it is automatically valid for any reference system, where the mathematics is more complex. We remember that, for geodesic coordinates, defined in a given point, the Christoffel symbols are locally null, at the same point. The equivalence principle, then, means that locally, we can not decide whether we have a non-rotating free-fall state, in the presence of a gravitational field, or uniform motion in its absence. A linear accelerated frame, relative to an inertial one, in the absence of gravitation, is locally indistinguishable from a rest-frame, immersed in a gravitational field.

Final Observations:

1st) Not all metric elements from Riemannian Geometry are adapted for representing the gravitational field of GRT; in fact we must have a "spacetime" metric, which is defined in order that, locally, it should reduce to Minkowskian's metric. Other arbitrary 4-dimensional Riemannian spaces can not represent spacetime, because of the requirement that GRT should reduce to Special Relativity.

2nd) The difference between GRT's spacetime fit to represent a gravitational field, and the one of Special Relativity, is that, in at least one infinitesimal neighborhood of a spacetime "event", the Ricci and Riemann tensors must be non-null.

3rd) In GRT, the energy-momentum tensor $T_{\mu\nu}$ must at least have one non-zero component in at least one infinitesimal neighborhood of a spacetime "event" or else, we shall be in a zero-gravitational case.

4th) The energy-momentum tensor employed in GRT must reduce to a symmetric tensor in Special Relativity.

Chapter 7

Robertson-Walker's Metric

7.1. Derivation of the Homogeneous and Isotropic Metric

We want to obtain a model of the Universe, capable of making possible a comparison between theoretical parameters and astronomical experimental data. It is necessary to use the argument of simplicity, because on observing the Universe as a whole, we find many "small" complexities, so that the big picture can only be obtained under simplifying hypotheses. The first one, which we now present, is known as the Cosmological Principle:

There are no privileged observers in the Universe.
All points and all directions are equivalent.

In other words, the Universe must be represented by a metric which, in its tridimensional space contents, is homogeneous and isotropic.

From the "simplicity" criterion stems the perfect fluid hypothesis, namely, that the matter that is contained in the Universe can be represented by the perfect fluid energy tensor. For a comoving observer, that, at a given cosmic time "t", is "at rest" in the given coordinate system:

$$T^{ij} = p\delta^{ij} \ ; \qquad\qquad (i, j = 1,2,3)$$

$$T^{i0} = T^{0i} = 0 \ ;$$

$$T^{00} = \rho,$$

where ρ and p stand for the energy density and cosmic pressure respectively. Hence, the comoving observer sees the fluid around him as isotropic, and the reference system is locally inertial. It has been verified (see section 7.4) that the Universe is not static, but in fact, it is expanding; in Special Relativity, the metric employed was Minkowski's. We now write it in a slightly different form, by including a scale factor $R(t)$ so that the differentials dx, dy, dz suffer a dilatation with increasing time:

$$ds^2 = dt^2 - R^2(t)[dx^2 + dy^2 + dz^2]. \tag{7.1.1}$$

As we know well, Minkowski's metric is a flat metric; the above metric has spatial part:

$$d\sigma^2 = R^2(t)[dx^2 + dy^2 + dz^2], \tag{7.1.2}$$

which is obviously also flat. In fact, this is a zero curvature submetric. Metric (7.1.1) is called Robertson-Walker's metric for a flat Universe.

For a three-dimensional space with curvature, the simplest case is that of "constant curvature". Its Riemann-Christoffel tensor is given by:

$$R_{abcd} = k(g_{ac}g_{bd} - g_{ad}g_{bc}), \tag{7.1.3}$$

where k stands for the tri-curvature. For the flat metric $k = 0$. (the reader can check this easily). Its Ricci tensor is then given by:

$$R_{\beta\delta} = g^{\alpha\gamma}R_{\alpha\beta\gamma\delta} = 2kg_{\beta\delta}. \tag{7.1.4}$$

In tridimensional space, like in the derivation we made for the spherically symmetric Schwarzschild's metric, we now expect, a spatial part of the metric in the form:

$$d\sigma^2 = e^{\alpha(r)}dr^2 + r^2(d\theta^2 + \sin^2\theta d\phi^2). \tag{7.1.5}$$

For this metric, we find:

$$R_{11} = \tfrac{\alpha'}{r}\ ; \qquad R_{22} = \operatorname{cosec}^2\theta\, R_{33} = 1 + \tfrac{1}{2}re^{-\alpha}\alpha' - e^{-\alpha}. \tag{7.1.6}$$

Then, the condition of constant space curvature reduces to the following equations:

$$\tfrac{\alpha'}{r} = 2ke^{\alpha}, \tag{7.1.7}$$

$$1 + \tfrac{1}{2}re^{-\alpha}\alpha' - e^{-\alpha} = 2kr^2. \tag{7.1.8}$$

The solution for the above is:

$$e^{-\alpha} = 1 - kr^2. \tag{7.1.9}$$

The expanding spatial part would now become:

$$d\sigma^2 = R^2(t)[\tfrac{dr^2}{1-kr^2} + r^2(d\theta^2 + \sin^2\theta d\phi^2)]. \tag{7.1.9a}$$

We now compare with the metric (7.1.2), which in spherical coordinates would have the form:

$$d\sigma^2 = R^2(t)[dr^2 + r^2(d\theta^2 + \sin^2\theta d\phi^2)]. \tag{7.1.10}$$

We said the this last one was a flat metric, which now is shown to belong to the case $k = 0$ (zero tricurvature). We are thus led to the Robertson-Walker's metric for the Universe,

$$ds^2 = dt^2 - R^2(t)[\tfrac{dr^2}{1-kr^2} + r^2(d\theta^2 + \sin^2\theta d\phi^2)]. \quad (7.1.11)$$

For this metric, the 3-dimensional curvature scalar is given by:

$$^3K(t) = kR^{-2}(t). \quad (7.1.12)$$

Basically, we have three possible cases in Cosmology: $k = 0$; $k > 0$; and $k < 0$, representing flat, positively curved or negatively curved spaces.

For practical reasons one can express Robertson-Walker's metric in a slightly different form:

$$ds^2 = dt^2 - \tfrac{R^2(t)}{[1+\frac{k\bar{r}^2}{4}]^2}[d\bar{r}^2 + \bar{r}^2(d\theta^2 + \sin^2\theta d\phi^2)], \quad (7.1.13)$$

where we made use of the substitution:

$$r = \tfrac{\bar{r}}{1+\frac{k}{4}\bar{r}^2}. \quad (7.1.14)$$

Usually, the bar in the "new" radial coordinate are dropped from the expression of the metric, and the new 3-curvature k' is also introduced, by means of a re-scaling:

$$k = |k|k', \quad (7.1.15)$$

so that, the three cases above are expressed by $k' = 0, +1$, or -1. We then rescale the radial coordinate, by:

$$\bar{\bar{r}} = \sqrt{|k|}\bar{r}, \quad (7.1.16)$$

and also we rescale the scale factor, by:

$$\bar{R}(t) = \tfrac{R(t)}{\sqrt{k}} \quad \text{for } k \neq 0,$$

or, (7.1.17)

$$\bar{R}(t) = R(t) \text{ for } k = 0.$$

We still keep the same form of the metric, on dropping bars again, so that Robertson-Walker's metric can either be expressed in each of these alternative forms:

$$ds^2 = dt^2 - R^2(t)[\tfrac{dr^2}{1-kr^2} + r^2(d\theta^2 + \sin^2\theta \; d\phi^2)], \quad (7.1.15)$$

or

$$ds^2 = dt^2 - \tfrac{R^2(t)}{[1+\frac{k}{4}r^2]^2}[dr^2 + r^2(d\theta^2 + \sin^2\theta \; d\phi^2)]. \quad (7.1.16)$$

The cosmological models which follow from this metric, in the case of null Cosmological Constant ($\Lambda = 0$), are called FRW's models (Friedmann - Robertson - Walker models); in the case of non zero Cosmological Constant ($\Lambda \neq 0$) they are called Lemaître's models.

7.2. Einstein's Field Equations for Cosmology

In the year 1922, the Russian cosmologist Friedmann, and then in 1927, the Belgian Abbé Lemaître, applied Einstein's field equations into the above metric.

Consider R.W's metric given by (7.1.15). On calculating the Christoffel symbols, we find the non-vanishing terms:

$$\Gamma^0_{11} = \frac{R\dot{R}}{1-kr^2}$$

$$\Gamma^0_{22} = R\dot{R}r^2$$

$$\Gamma^0_{33} = R\dot{R}r^2 \sin^2\theta \tag{7.2.1}$$

$$\Gamma^1_{01} = \frac{\dot{R}}{R}$$

$$\Gamma^1_{11} = \frac{kr}{1-kr^2}$$

$$\Gamma^1_{22} = -r(1-kr^2)$$

$$\Gamma^1_{33} = -r(1-kr^2)\sin^2\theta$$

$$\Gamma^2_{02} = \frac{\dot{R}}{R}$$

$$\Gamma^2_{12} = \frac{1}{r}$$

$$\Gamma^2_{33} = -\sin\theta\cos\theta$$

$$\Gamma^3_{03} = \frac{\dot{R}}{R}$$

$$\Gamma^3_{13} = \frac{1}{r}$$

$$\Gamma^3_{23} = \cot\theta.$$

We now write the Ricci tensor, in the form:

$$R_{\mu\nu} = \Gamma^\sigma_{\mu\sigma,\nu} - \Gamma^\sigma_{\mu\nu,\sigma} + \Gamma^\rho_{\mu\sigma}\Gamma^\sigma_{\rho\nu} - \Gamma^\rho_{\mu\nu}\Gamma^\sigma_{\rho\sigma}. \tag{7.2.2}$$

Applying the above formulae, we find the non-vanishing terms:

$$R_{00} = \frac{3\ddot{R}}{R}\ ;$$

$$R_{11} = -(R\ddot{R} + 2\dot{R}^2 + 2k)/(1 - kr^2);$$
$$R_{22} = -(R\ddot{R} + 2\dot{R}^2 + 2k)r^2; \qquad (7.2.3)$$
$$R_{33} = R_{22}\sin^2\theta \quad .$$

On the other hand, the perfect fluid energy-momentum tensor is given by:

$$T^{\mu}_{\nu} = (\rho + p)u^{\mu}u_{\nu} - p\delta^{\mu}_{\nu}. \qquad (7.2.4)$$

We remember that the 4-velocity obeys the relation:

$$u^{\mu}u_{\mu} = 1, \qquad (7.2.5)$$

where $u^{\mu} \equiv \frac{dx^{\mu}}{ds}$, so that:

$$T \equiv T^{\mu}_{\mu} = (\rho + p) - 4p = \rho - 3p. \qquad (7.2.6\ )$$

In the comoving reference system:

$$u^{\mu} = \delta^{\mu}_{0}, \qquad (7.2.7)$$

$$u_{\mu} = g_{\mu\nu}\delta^{\nu}_{0} = g_{\mu 0} = \delta^{0}_{\mu}, \qquad (\ 7.2.8)$$

$$T_{\mu\nu} = (\rho + p)\delta^{0}_{\mu}\delta^{0}_{\nu} - pg_{\mu\nu}\ , \qquad (7.2.9)$$

$$T_{\mu\nu} - \tfrac{1}{2}Tg_{\mu\nu} = (\rho + p)\delta^{0}_{\mu}\delta^{0}_{\nu} - pg_{\mu\nu} - \tfrac{1}{2}(\rho - 3p)g_{\mu\nu} = (\rho + p)\delta^{0}_{\mu}\delta^{0}_{\nu} - \tfrac{1}{2}(\rho - p)g_{\mu\nu}. \quad (7.2.10)$$

We find now:

$$T_{00} - \tfrac{1}{2}Tg_{00} = \tfrac{1}{2}(\rho + 3p), \qquad (7.2.11)$$

$$T_{11} - \tfrac{1}{2}Tg_{11} = \frac{(\rho - p)R^2}{2(1 - kr^2)}\ , \qquad (7.2.12)$$

$$T_{22} - \tfrac{1}{2}Tg_{22} = \tfrac{1}{2}(\rho - p)R^2r^2\ , \qquad (7.2.13)$$

$$T_{33} - \tfrac{1}{2}Tg_{33} = \tfrac{1}{2}(\rho - p)R^2r^2\sin^2\theta, \qquad (7.2.14)$$

while the non-diagonal terms vanish.

The reader must become puzzled from the calculations made in (7.2.11) to (7.2.14). The explanation is given now. From Einstein's field equations:

$$G^{\mu\nu} \equiv R^{\mu\nu} - \tfrac{1}{2}Rg^{\mu\nu} = -\kappa T^{\mu\nu}, \qquad (7.2.15)$$

we can derive the alternative expression,

$$R^{\mu\nu} = -\kappa\left(T^{\mu\nu} - \tfrac{1}{2}Tg^{\mu\nu}\right). \qquad (7.2.16)$$

Hint:
Contract the mixed form of equation (7.2.15), $R^{\mu}_{\nu}-\frac{1}{2}R\delta^{\mu}_{\nu}=-\kappa T^{\mu}_{\nu}$; show that $R=-\kappa T$; and then verify the required result.

Let us return to Einstein's field equations; we are left with two equations:

$$\frac{3\ddot{R}}{R}=-\frac{1}{2}\kappa(\rho+3p)\,, \tag{7.2.17}$$

$$R\ddot{R}+2\dot{R}^2+2k=\frac{1}{2}\kappa(\rho-p)R^2. \tag{7.2.18}$$

It is usual to write one equation for ρ alone, and other for p alone:

$$\kappa\rho=3\frac{\dot{R}^2}{R^2}+\frac{3k}{R^2}, \tag{7.2.19}$$
$$\kappa p=-\frac{2\ddot{R}}{R}+\frac{\dot{R}^2}{R^2}-\frac{k}{R^2}.$$

For the Lemaître models we add a density term $\frac{\Lambda}{\kappa}$, while we add a negative cosmic pressure $-\frac{\Lambda}{\kappa}$ thus finding:

$$\kappa\rho=3H^2+\frac{3k}{R^2}+\Lambda, \tag{7.2.20}$$

$$\kappa p=-\frac{2\ddot{R}}{R}-H^2-\frac{k}{R^2}-\Lambda. \tag{7.2.21}$$

Note that we have defined Hubble's parameter, $H\equiv\frac{\dot{R}}{R}$, which determines the rate of expansion of the Universe. Notice also that the inclusion of the Λ constant as we did above, is justified by the results of section 3.5. Zeldovich calculated, from Quantum mechanical arguments, further that this procedure is equivalent to attributing an inherent density and negative pressure to the vacuum. This interpretation comes from displacing the Λ term from the left-hand side of the field equations to the r.h.s. In the l.h.s situation:

$$G_{\mu\nu}+\Lambda g_{\mu\nu}=-\kappa T_{\mu\nu},$$

we say that Λ is of geometric origin; however, if you write in the r.h.s.:

$$G_{\mu\nu}=-\kappa T_{\mu\nu}-\Lambda g_{\mu\nu}=-\kappa\overline{T}_{\mu\nu},$$

where we have defined an augmented energy tensor, which means that Λ is included in the new energy-tensor, as a property of matter, and not of the geometry. In both situations, we have the conservation of energy tensor:

$$T^{\mu}_{\nu;\mu}\equiv\overline{T}^{\mu}_{\nu;\mu}=0.$$

Likewise, we have for the l.h.s form:

$$[G^{\mu}_{\nu} + \Lambda g^{\mu}_{\nu}]_{;\mu} \equiv G^{\mu}_{\nu;\mu} = 0,$$

so that both forms are conserved.

From equations (7.2.20) and (7.2.21) we obtain:

$$\dot{\rho} + 3H(\rho + p) = 0. \tag{7.2.22}$$

The above equation could also have been derived by imposing the conservation of the energy momentum tensor:

$$T^{\mu\nu}_{;\nu} = 0. \tag{7.2.23}$$

In fact, we get two equations:

$$(\rho u^{\mu})_{;\mu} + p u^{\mu}_{;\mu} = 0, \tag{7.2.24}$$

and

$$(\rho + p) u^{\nu}_{;\mu} u^{\mu} = (g^{\mu\nu} - u^{\mu} u^{\nu}) p_{,\mu}. \tag{7.2.25}$$

From (7.2.24),

$$\rho_{,\mu} u^{\mu} + (\rho + p)(u^{\mu}_{,\mu} + \Gamma^{\mu}_{\nu\mu} u^{\nu}) = 0. \tag{7.2.26}$$

If we now plug (7.2.7) into (7.2.26) we obtain (7.2.22), which is also called the continuity equation.

Now let us look to (7.2.25). For a comoving observer, we have:

$$u^{\mu} = \delta^{\mu}_{0}. \tag{7.2.7}$$

With $\mu \neq 0$, we have:

$$(\rho + p) u^{\nu}_{;\mu} u^{\mu} = 0,$$

because Krönecker's delta equals zero for $\mu \neq 0$, while:

$$\frac{\partial p}{\partial r} \equiv \frac{\partial p}{\partial \theta} \equiv \frac{\partial p}{\partial \phi} \equiv 0\,,$$

because there is no pressure gradient, and the energy density and cosmic pressure only depend on time, due to homogeneity and isotropy. Hence,

$$\rho = \rho(t)\,;$$

$$p = p(t)\,; \quad \text{and}$$

$$(g^{\mu\nu} - \mu^{\mu}\mu^{\nu}) p_{,\mu} = 0$$

with $\mu \neq 0$.

For $\mu = 0$, as the metric tensor has zero covariant derivative, the l.h.s. of the (7.2.25) is equal to zero, leaving us with:

$$(g^{0\nu} - \mu^0\mu^\nu)p_{,0} = 0.$$

Because in general, $\frac{\partial p}{\partial t} \neq 0$, we have the above relation reduced to:

$$\delta^0_\nu - \delta^0_\nu = 0.$$

On the other hand, we have already seen that the relation:

$$\mu^\nu_{;\mu}\mu^\mu = 0,$$

represents a geodesic, so that the fluid particles move along them; this is expected, indeed, since the particles are not subject to a pressure gradient to push them off geodesics $(\nabla p = 0)$.

A useful identity may be derived by calculating the time derivative of the expression:

$$\dot{R}(t) \equiv H(t)R(t).$$

Then we find, by the product rule:

$$\ddot{R}(t) \equiv \dot{H}(t)R(t) + H(t)\dot{R}(t), \text{ or,}$$

$$\frac{\ddot{R}(t)}{R} \equiv \dot{H}(t) + H^2(t).$$

7.3. Constant Deceleration Parameter Models

As we have mention before, we define Hubble's parameter by:

$$H = \frac{\dot{R}}{R}.$$

Berman (1983) proposed that H varied with a power of $R = R(t)$, such as:

$$H = DR^{-m}, \tag{7.3.1}$$

with $D, m =$ constants.

For $m = 0$ we find H constant and by integration we obtain the law of variation of the scale-factor with time:

$$R = R_0 e^{Dt}, \tag{7.3.2}$$

with $R_0 =$ constant.

We find such law when dealing with exponential inflation, to be considered later.

If $m \neq 0$, we find, by integration, a possible law:

$$R = R(t) = (mDt)^{1/m}, \tag{7.3.3}$$

while from Einstein's field equations with $\wedge \neq 0$, we find:

$$\rho = \frac{3D^2}{\kappa}R^{-2m} + \frac{3}{\kappa}kR^{-2} + \frac{\wedge}{\kappa}, \tag{7.3.4}$$

and,

$$p = \frac{(2m-3)D^2}{\kappa}R^{-2m} - \frac{k}{\kappa}R^{-2} - \frac{\wedge}{\kappa}. \tag{7.3.5}$$

Berman (1983), and Berman and Som (1989) found that these models have constant deceleration parameter $q = m - 1 = \text{constant}$, where we define:

$$q = -\frac{R\ddot{R}}{\dot{R}^2}. \tag{7.3.6}$$

The deceleration parameter was believed to be positive for the present Universe, that is why it was defined negatively; it is now thought that the present Universe is in fact in a process of acceleration, as we shall see later. An alternative definition, completely equivalent is:

$$q = -\frac{\ddot{R}}{RH^2}. \tag{7.3.7}$$

In order to show that the deceleration parameter is constant for the power-law (7.3.3) we take derivatives of $R(t)$ finding:

$$\dot{R} = DR^{-m+1}, \tag{7.3.8}$$

and,

$$\ddot{R} = -D^2(m-1)R^{-2m+1}. \tag{7.3.9}$$

We now plug (7.3.8) and (7.3.9) into (7.3.6) finding:

$$q = m - 1 = const. \tag{7.3.10}$$

We now obtain for Hubble's parameter the following law:

$$H = (mt)^{-1}. \tag{7.3.11}$$

The present author has been apparently the first one to publish the above formula, or its equivalent form:

$$H = [(1+q)t]^{-1}. \tag{7.3.12}$$

In other words, the age of the Universe would be given by $\frac{H^{-1}}{m} = \frac{H^{-1}}{q+1}$. Astronomers have found that:

$$H_0^{-1} \cong 1.8\text{x}10^{10} years, \tag{7.3.13}$$

where the subscript zero means "present value".

If we now calculate $\dddot{R}$, we find:

$$\dddot{R} = -D^2(m-1)(1-2m)\dot{R}R^{-2m}, \tag{7.3.14}$$

so that the variation of deceleration parameter is given by the pure number (adimensional number), suggested by Peter Landsberg(1983), who also made references to q – constant models in terms of rough calculations:

$$Q = \frac{R^2\dddot{R}}{\dot{R}^3} = (2m-1)(m-1). \tag{7.3.15}$$

From (7.3.15) with (7.3.9) we find:

$$Q = 2q^2 + q. \tag{7.3.16}$$

We conclude that the Q–parameter is also constant, for constant deceleration parameter models.

By plugging (7.3.8) and (7.3.9) into (7.3.4) and (7.3.5), we can find the time dependence for the energy density and cosmic pressure:

$$\rho = \left(\frac{3}{\kappa m^2}\right)t^{-2} + \frac{3k}{\kappa(m^2D^2)^{1/m}}t^{-2/m} + \frac{\Lambda}{\kappa}, \tag{7.3.17}$$

and,

$$p = \left[\frac{(2m-3)}{\kappa m^2}\right]t^{-2} - \frac{k}{\kappa(m^2D^2)^{1/m}}t^{-2/m} - \frac{\Lambda}{\kappa}. \tag{7.3.18}$$

It can be checked that $k = 0$ and $m = \frac{3}{2}$ and $\Lambda = 0$ gives the so called "Friedmann flat model" where $\rho = \rho_{critical} = 3H^2/\kappa$. This critical density is one of the most important parameters in Cosmology (See Section 7.7).

7.4. Cosmological Red-Shift

We shall now show that, for a light wave emitted in cosmic time t_E, when the scale-factor is $R(t_E)$, and which is received at present time t_0, when the scale-factor is $R(t_0) > R(t_E)$ – (in other words, the Universe is expanding) – the wavelength λ_0 received is larger than the one emitted, λ_E :

$$\frac{\lambda_0}{\lambda_E} = \frac{R(t_0)}{R(t_E)} > 1. \tag{7.4.1}$$

The spectrum of such light shows a RED-SHIFT.

Consider a wave front of an electromagnetic emission at the initial time t_E. For a radial geodesic,

$$d\theta = d\phi = 0.$$

The radial coordinate being "r" for the emitting "galaxy", we have from Robertson-Walker's metric, with the $ds = 0$,

$$-\frac{R^2(t)}{(1+\frac{k}{4}r^2)^2}dr^2 + dt^2 = 0. \qquad (7.4.2)$$

We expect for a far-away galaxy, that the periods dt_E, and dt_0, of the emitted and received photon, are much smaller than the total time of the trajectory, $[t_0 - t_E]$. From equation (7.4.2), we find now that:

$$\int_{t_E}^{t_0} \frac{dt}{R(t)} = -\int_r^0 \frac{dr}{1+\frac{kr^2}{4}}, \qquad (7.4.3)$$

and also,

$$\int_{t_E+dt_E}^{t_0+dt_0} \frac{dt}{R(t)} = -\int_r^0 \frac{dr}{1+\frac{kr^2}{4}}. \qquad (7.4.4)$$

We now recall the linearity of the definite integral operation:

$$\int_{t_E+dt_E}^{t_0+dt_0} \equiv \int_{t_E+dt_0}^{t_E} + \int_{t_E}^{t_0} + \int_{t_0}^{t_0+dt_0}, \qquad (7.4.5)$$

so that,

$$\int_{t_E+dt_E}^{t_0+dt_0} \frac{dt}{R(t)} = -\frac{dt_E}{R(t_E)} + \int_{t_E}^{t_0} \frac{dt}{R(t)} + \frac{dt_0}{R(t_0)} = \int_{t_E}^{t_0} \frac{dt}{R(t)}, \qquad (7.4.6)$$

and, taking this result into (7.4.3) and (7.4.4) we find:

$$-\frac{dt_E}{R(t_E)} + \frac{dt_0}{R(t_0)} = 0, \qquad (7.4.8)$$

which takes us back to (7.4.1), where we have already included the result that we know that $R(t_0) > R(t_E)$, because:

$$1 + \left(\frac{\Delta\lambda}{\lambda_E}\right) \equiv 1 + z = \frac{R_0}{R_E} = \frac{R(t_0)}{R(t_E)} > 1. \qquad (7.4.9)$$

Notice that $z > 0$, for a red-shift ($\Delta\lambda > 0$) and that we have used the standard equation:

$$\lambda = cT,$$

where T stands for period, and λ is the wavelength.

It was the American astronomer Edwin Hubble, the first, or one of the first, to discover the expansion of the Universe by analyzing the spectra of light from nearby galaxies. He found Hubble's law, under which, for nearby galaxies,

$$z \cong H_0 d_L, \qquad (7.4.10)$$

where $H_0 = H(t_0) =$ Hubble's "constant", and d_L stands for the "luminosity" distance of the galaxy as seen from the earth. In next section, we shall show what we mean by d_L which is roughly the distance, from the earth, of the galaxy; while we shall also show how to obtain Hubble's law from Roberston-Walker's metric.

7.5. Hubble's Law

If the physical distance is "d" between two galaxies, which are comoving, and are separated by coordinate distance d_0, we find:

$$d = R(t)d_0 = d(t). \tag{7.5.1}$$

In fact, the above is the translation of what we mean by the expanding metric.

The relative speed between both galaxies, is, then:

$$v = \frac{d}{dt}[d(t)] = \dot{R}d_0 = \dot{R}[\frac{d(t)}{R(t)}] = H(t)d(t). \tag{7.5.2}$$

This speed is interpreted, by forgetting peculiar random speeds of galaxies, as a recessional speed. So, Hubble measured red-shifts and attributed them to recessional speeds, that are as much larger as its luminosity distance is from us.

For a comoving source and observer, with distance d apart, with $d << 1$, we know that:

$$1 + z \cong \frac{R_0}{R_E}. \tag{7.5.3}$$

We now expand R_0 about R_E keeping up to the first derivative in Taylor's series:

$$R_E \cong R_0 - \dot{R}_0[\Delta t_1], \tag{7.5.4}$$

where,

$$\Delta t_1 \equiv t_0 - t_1 \equiv \frac{d}{c}, \tag{7.5.5}$$

and where d is loosely called the distance between the two galaxies. Then,

$$z = \frac{R_0}{R_E} - 1 \cong \frac{1}{1 - H_0 d} - 1 \cong H_0 d \qquad \text{q.e.d.}$$

7.6. Cosmological Model for Large Lambda

Let us recall the two equations for Cosmology, one for energy density, the other for cosmic pressure (7.2.20) and (7.2.21):

$$\kappa\rho = 3H^2 + \frac{3k}{R^2} + \Lambda, \tag{7.6.1}$$

$$\kappa p = -\frac{2\ddot{R}}{R} - H^2 - \frac{k}{R^2} - \Lambda. \tag{7.6.2}$$

Suppose that Λ is the driving "force" in the Universe, so that:

$$\kappa\rho << \Lambda, \tag{7.6.3}$$

$$\Lambda >> R^{-2} > 0, \tag{7.6.4}$$

$$\Lambda >> \kappa p. \tag{7.6.5}$$

Note that in the ρ and p, terms, of this Section, the Λ – contribution is not present; what they represent, is just the real matter fields.

From (7.6.1) we find:

$$3H^2 \cong \Lambda, \tag{7.6.6}$$

or,

$$\frac{\dot{R}}{R} \cong \sqrt{\frac{\Lambda}{3}}. \tag{7.6.7}$$

On integrating,

$$R \cong R_0 e^{\sqrt{\frac{\Lambda}{3}}\,t}, \tag{7.6.8}$$

where $R_0 = R(t = 0) = $ constant.

Equation (7.6.2) is then fulfilled identically. This scale-factor is representative of exponential inflation (see next Section). From next Chapter, we shall see that for Planck's Universe, in the very early Universe, which stands at the limit between the validity of Classical Physics and Quantum Physics, Λ is in fact the largest driving cause for the expansion of the Universe. From Einstein's equations as above, we can predict thus, that inflationary expansion must occur in the very early Universe. It could be argued that, because for the present Universe, the Λ constant is responsible for a very large amount of the energy density of the present Universe ($\frac{\Lambda}{\kappa}$ is very large), the present Universe could be also in an exponential inflationary phase. For this phase, the deceleration parameter is $q = -1$, as can be checked by plugging the inflationary law into the definition for q (see Section 7.3). In fact, we can imagine that $-1 \leq q < 0$. So that, the Universe is accelerating, as recent Supernovae observations have revealed. However, it seems that $\frac{\Lambda}{\kappa}$ is not the only driving "force" in the present Universe (that is why q is not equal to -1 ; by the same token, we could have, or not, the equation of state, $p \neq -\frac{1}{3}\rho$. (See Section 9.7).

It seems however, that the Λ value for the very early Universe should be 10^{120} times larger than its value for the present Universe. The Λ value today is limited by the fact that Schwarzschild's solution for the solar system is more or less accurate with $\Lambda \approx 0$, according to astronomers.

We mention here that the Schwarzschild's metric with a Λ constant included is given by (Tolman, 1934):

$$ds^2 = \frac{-dr^2}{1-\frac{2m}{r}-\frac{\Lambda}{3}r^2} - r^2(d\theta^2 + \sin^2\theta d\phi^2) + c^2\left(1 - \frac{2m}{r} - \frac{\Lambda}{3}r^2\right)dt^2. \tag{7.6.9}$$

7.7. Inflation

A closely related idea, to the large lambda model considered above, is "inflation". As we shall see in next Section, and further in next Chapter, one is led to believe that, at least, for the very early Universe, an exponential scale-factor driven by a large lambda, has in fact occurred. From Grand Unified Theories (GUT's) which tell that the unified field covered by the nuclear strong and weak, and the electromagnetic, interactions, had "afterwards", under the action of a scalar field, broken its "symmetry". Then, the energy of the vacuum (V_0) was driving the Universe, say with $\rho \cong V_0$. This is tantamount to choosing a model with large Λ, where according to last Section,

$$H = \sqrt{\frac{\kappa}{3} V_0} = \sqrt{\frac{\Lambda}{3}} \, .$$

GUT's time, according to Particle Physics is about 10^{-38} seconds after the "birth" of the Universe ($t = 0$), which is very close to Planck's time, 10^{-43} seconds, which is the minimum time for the failure of Classical Physics and the entrance of Quantum Physics in the scenario (see next Section), when we go back in time.

Among the problems solved by inflation we list the "flatness" and the "horizon" problems plus "monopoles" production in the Universe. The "flatness" problem is basically how one can explain that the Universe today is so close to being flat. For instance, the visible matter energy density, would point to a magnitude of the energy density only one hundred times less the critical density for the Universe, which is a "close" number. The critical density is the density for which $k = 0$ in the Robertson-Walker's metric when the Universe is pressureless (Friedmann models). The density parameter is defined for the total energy density as $\Omega = \frac{\kappa\rho}{3H^2}$, so that for the critical density, $\Omega = 1$. Analogously, we define for the cosmological constant, a density parameter, $\Omega_\Lambda = \frac{\Lambda}{3H^2}$, while for the energy density of matter we define $\Omega_m = \frac{\kappa\rho_m}{3H^2}$, where ρ_m denotes the energy density of the matter in the Universe. It should be noted that experimentally $\Omega_\Lambda \sim 0.7$ and then if $k = 0$, we shall have $\Omega_m \sim 0.3$. From (7.6.1), it follows that the energy density is the "critical" one, when $k = \Lambda = 0$.

Returning to the flatness problem, this is caused because it would have required a very fine tuning of the initial conditions for the Universe in order to obtain, today, a flat Universe.

The horizon problem is defined by the question of how can we explain why different disconnected parts of the Universe have similar or same physical properties. This could not have been achieved by power law scale factors, that would expand faster than the speed of light traveling from the one part to the other in the Universe, sufficiently far away. The rapidly exponential growth solves this problem, because it allows the early radius of space to be much smaller than the horizon distance $r_h = ct$.

The flatness problem is also solved by the rapid expansion caused by an inflationary phase because it turned the Universe more flat quickly. The best picture for this phenomenon is to imagine the Universe as the surface of a tridimensional balloon: as the balloon inflates the surface becomes more flat.

The monopoles production in the very early Universe is a problem, for no trace of such monopoles, required by Particle Physics theories to have been produced in the very early Universe, are found in the present Universe. The explanation given by inflationary theorists is that there was a finite amount of monopoles before inflation; with the rapidly expansion caused by inflation, they were diluted, so that no trace is found today.

7.8. Planck's Universe and the Λ-problem

The limit for Classical Cosmology becomes evident when one tries to combine the constants of Physics into dimensional quantities. For Classical Physics the constants are two, namely, the velocity of light in vacuum, c, and Newton's gravitational constant, G: one represents electromagnetism, and the other, the gravitation. For Quantum Physics there is another constant, (Planck's), $\hbar = h/2\pi$.

When we combine all of them, we obtain a length dimension, the Planck's length:

$$L_{Pl} = (\frac{\hbar G}{c^3})^{\frac{1}{2}} \cong 1.6\text{x}10^{-33} cm.$$

Analogously, we find Planck's time (t_{Pl}) and Planck's mass (M_{Pl}):

$$t_{Pl} = (\frac{\hbar G}{c^5})^{\frac{1}{2}} \cong 5\text{x}10^{-44} \sec,$$
$$M_{Pl} = (\frac{\hbar c}{G})^{\frac{1}{2}} \cong 2\text{x}10^{-5} g\ .$$

We could try to find a value for the Λ - constant, by constructing a combination of the constants yielding a length to the minus second power. This is:

$$\Lambda_{Pl} \cong L_{Pl}^{-2} \cong 1\text{x}10^{66} cm^{-2},$$

regarded as the Cosmological constant for Planck's Universe.

On the other hand, though we can keep the above values, for Planck's Universe, which represents the limit between Quantum gravity theory and Classical General Relativity, we know that the above value for Λ_{Pl} can not exist for the present Universe. In fact, it is estimated from supernovae high red-shift observations that:

$$\Lambda_0 \cong 10^{-56} cm^{-2}.$$

The above experimental result, matches with the upper limit for the "constant", which could possibly enter in the Schwarzschild's metric with Λ (see equation 7.6.9), without disturbing the data gathered from solar system's observations, when compared with the same equation with $\Lambda = 0$.

The ratio:

$$\Lambda_{Pl}/\Lambda_0 \approx \frac{10^{66}}{10^{-56}} = 10^{122}\ ,$$

constitutes the most serious problem of contemporary Cosmology, being called the Cosmological Constant Problem or briefly the lambda-problem.

In the next Chapters we shall point out the solution for this problem, according to several authors, with whom we are included. We refer to Part IV for further information.

7.9. A White-Hole-Robertson-Walker's Universe

The contents of this Section, will be published soon by the journal "Astrophysics and Space Science" (Berman, 2006h), and is based on a method first published by Gomide and Berman (1987).

Pathria(1972) published a paper showing that, for certain values of the cosmological constant, a pressureless finite (positively curved) Universe could be inside a white-hole, if it obeyed Robertson-Walker's metric (with $k = 1$):

$$ds^2 = dt^2 - \frac{R^2(t)}{\left[1+\left(\frac{kr^2}{4}\right)\right]^2} d\sigma^2, \tag{7.9.1}$$

where,

$$d\sigma^2 = dx^2 + dy^2 + dz^2. \tag{7.9.2}$$

If R stands for the radius of a "large sphere" of mass M, while G stands for Newton's gravitational constant, a white-hole could be described by the condition that R is smaller than its Schwarzschild's radius:

$$R < 2GM. \tag{7.9.3}$$

It is known, on experimental grounds, that the present Universe, obeys the Machian relation by Brans and Dicke(1961),

$$R \sim GM. \tag{7.9.4}$$

We, thus, can think that the Universe may be inside a white-hole, but this should be analyzed by means of Einstein's field equations for the above metric,

$$\kappa\rho = 3H^2 + \frac{3k}{R^2} + \Lambda, \tag{7.9.5}$$

$$\kappa p = -2\frac{\ddot{R}}{R} - H^2 - \frac{k}{R^2} - \Lambda, \tag{7.9.6}$$

where,

$$H = \frac{\dot{R}}{R},$$

and,

$$\kappa = 8\pi G.$$

We study below, all three tricurvature cases ($k = 0, \pm 1$). We also treat the absolute temperature of black holes and the Universe.

Berman(2006;2006a) has studied the same subject by means of Mach's Principle, and the hypothesis of a zero-total-energy for the Universe. (See Chapter 9, and Section 10.6)

Closed Universe as a White-hole

Considering a positive cosmological constant ($\Lambda > 0$), and remembering that the volume V, for a closed Robertson-Walker's Universe, is given by(Adler, Bazin and Schiffer, 1975):

$$V = 2\pi^2 R^3, \tag{7.9.7}$$

we find,

$$\rho = \frac{M}{2\pi^2 R^3}. \tag{7.9.8}$$

From equation(7.9.5) we find:

$$\kappa\rho = \frac{3\left[1+\dot{R}^2\right]}{R^2} + \Lambda = \frac{\kappa M}{2\pi^2 R^3} = \frac{2r}{\pi R^3}, \tag{7.9.9}$$

where r stands for Schwarzschild's radius, i.e.,

$$r = 2GM = R\left\{\frac{\pi}{2}\left[3\dot{R}^2 + 3 + \Lambda R^2\right]\right\} > R. \tag{7.9.10}$$

Because $\dot{R}^2 > 0$ because $R(t)$ is a real number, we find,

$$\Lambda > \left[\frac{2}{\pi} - 3\right] R^{-2} > -3R^{-2}. \tag{7.9.11}$$

As the scale-factor is essentially positive, condition (7.9.11) will be certainly fulfilled by the condition, $\Lambda > 0$.

We have shown that a closed Universe, while cosmic pressure p follows Einstein's field equation (7.9.6) normally, shall be inside a white-hole, if $\Lambda > 0$.

Flat Universe as a White-hole

We shall now apply Einstein's field equations for $k = 0$. With the prescription:

$$V = \frac{4}{3}\pi R^3, \tag{7.9.12}$$

so that,

$$\rho = \frac{3M}{4\pi R^3}, \tag{7.9.13}$$

we would find,

$$r = R^3\left[H^2 + \frac{\Lambda}{3}\right] = R\left[\dot{R}^2 + \frac{\Lambda}{3}R^2\right]. \tag{7.9.14}$$

We find that a sufficient condition for $r > R$ is,

$$\Lambda > \frac{3}{R^2}. \tag{7.9.15}$$

We remark that this condition is equivalent to "closing" the Universe, in what respects the density field equation (but not in terms of the pressure equation). Even if this condition may not prevail for the very early Universe, nevertheless, with increasing values of the scale-factor, eventually the condition (7.9.15) shall be fulfilled.

The limiting scale-factor, is of course,

$$R > R_0 = \sqrt{\frac{3}{\Lambda}}. \tag{7.9.16}$$

Absolute temperature in Cosmology and White-holes

Hawking(2001), cites the formula he obtained many years ago, for the absolute temperature T associated with a non rotating and non charged black-hole; though in our paper we are dealing with Classical Physics, and Hawking's formula is based on Quantum theory, it will be seen that we can make a bridge with Classical Cosmology:

$$T = \frac{\bar{h}\,c^3}{8\pi k R_S}, \ (\bar{h} = \frac{h}{2\pi}\ ) \tag{7.9.17}$$

where, k, and h stand respectively for Boltzman and Planck constants, and R_S stands for Schwarzschild's radius, in such a way that, if we associate the scale-factor with R_S, then, from (7.9.17),

$$T \cong \frac{hc^3}{16\pi^2 k} R^{-1}. \tag{7.9.18}$$

We now shall sketch the derivation of the above formula. Suppose that a virtual pair of photons is created in the neighborhood of event-horizon surface of a static Schwarzschild's black hole. If the negative energy photon crosses the surface into the black hole, the other one will be "emitted" from the surface towards the exterior space. The latter is described as radiation from the black hole. Suppose that each of the photons had energy $\pm\, pc$, where p stands for linear momentum. The negative energy photon will be trapped inside the black hole, so that the uncertainty in its position is of the order of Schwarzschild's radius, 2 R_S. The momentum uncertainty is of order p; if we equate the energy with the termal energy kT, where k stands for Boltzman's constant, T is the absolute temperature, and we make use of Heisemberg's uncertainty principle, we find:

$$\Delta p \approx \frac{\bar{h}}{2R_S} \approx \frac{kT}{c}\ , \text{ or,}$$

$$T \approx \frac{\bar{h}c}{2kR_S} \approx \frac{\bar{h}c^3}{2kGM}.$$

This is almost the same formula as (7.9.17).

If Brans-Dicke relation is taken, in such a way that $R \propto M$, we would have, from (7.9.17),

$$T \cong \frac{hc^3}{16\pi^2 k} R^{-1}. \tag{7.9.18}$$

This is exactly what we find in standard Cosmology, between the scale factor $R(t)$ and T. This is a most striking analogy that points out to a possible white-hole Universe. [See however, Section 10.6, where it is studied the Machian case $R \propto T^{-2}$].

Open Universe as a white-hole

We shall leave, once more, the cosmic pressure to obey freely Einstein's field equations. From the energy density ρ – equation (7.9.5), with $k = -1$, we have, then,

$$r = R\left[\frac{\alpha}{4\pi}\left(3R^2H^2 + \Lambda R^2 - 3\right)\right], \tag{7.9.19}$$

where we suppose, quite generally, that the tridimensional volume is given by:

$$V = \alpha R^3 \qquad (\alpha = \text{positive constant}). \tag{7.9.20}$$

The condition for the open Universe to be or become a white-hole, is given by $r > R$, so that,

$$\Lambda > \left[m - 3\dot{R}^2\right] R^{-2}, \tag{7.9.21}$$

where,

$$m = 3 + \frac{4\pi}{\alpha} > 3. \tag{7.9.22}$$

We conclude that the condition (7.9.21) can be staten as,

$$\Lambda > \left[3(1 - \dot{R}^2)\right] R^{-2}. \tag{7.9.23}$$

At least, we can impose the following sufficient condition:

$$\Lambda > 3R^{-2} > 0, \tag{7.9.24}$$

because $\dot{R}^2 > 0$.

By the same token as above, for the expanding Universe, when the scale factor becomes larger than $\sqrt{\frac{3}{\Lambda}}$ we shall have a white-hole-Universe.

We remark that condition (7.9.24) is identical with (7.9.16), which refers to a flat Robertson-Walker's metric.

Conclusions

We have shown, in a different context than in Pathria's paper (where $p = 0$, and Λ obeys certain conditions), that the closed Robertson-Walker's Universe, with any value of p constrained to obey Einstein's field equations may be thought as being a white-hole. Brans-Dicke relation to the problem is not conclusive, because it represents only a Machian condition for the Universe. In a similar way, flat or open Universes, in the expanding phase, may become white-holes after R becoming larger than $\sqrt{\frac{3}{\Lambda}}$. Finally, we motivate the analogy between the Universe and a white-hole, by means of their absolute temperatures. The Machian condition for the Universe, was taken, partially, as implying that the absolute temperature ran like $\left(\sqrt{R}\right)^{-1}$. (cf Section 10.6). The standard condition, however, is that $T \propto R^{-1}$, like we considered above.

Part IV

Energy of the Vacuum and the Λ – Universe

Chapter 8

The Λ Problem and Its Solution

8.1. Introducing the Λ- Problem

We have seen in the last Chapter that we have two distinct evaluations for the Λ constant: the first is the very early Universe Λ estimated as the second minus power of the Planck's length; the second is the estimated value by means of high red-shift supernovae observations, which is 10^{-120} times smaller, for the present Universe. This is really a problem because the "constant" is not a constant!!!. A possible solution for this enigma, as advanced by several authors including the present author (Berman and Som, 1990 a; Berman, 1990; 1991 a; 1991 b; 1992 b ;1992; 1994; Berman and Trevisan, 2002 a; 2001 a; 2001 b; 2001 c; Berman, Som and Gomide, 1989), following a hint by a physicist Orfeu Bertolami (1986), was to consider that we have not a constant, but really a Λ-term varying with a minus square power of the age of the Universe:

$\Lambda \propto t^{-2}$ (Bertolami-Berman solution for Brans-Dicke theory).

In fact,

$$\left(\frac{t_{GUT}}{t_0}\right)^{-2} \cong \left(\frac{10^{-34}\,\text{sec}}{10^{10} years}\right)^{-2} \cong 10^{120}.$$

This resembles approximately the necessary ratio. On the other hand, another suggested Λ – term variation was given by Chen and Wu (1990):

$$\Lambda \propto R^{-2}.$$

In fact, consider the ratio:

$$\left(\frac{L_{Pl}}{R_0}\right)^{-2} \cong \left(\frac{10^{-33} cm}{10^{28} cm}\right)^{-2} \cong 10^{122}.$$

An exhaustive paper, containing models with a Λ – term, with different kinds of dependence including power-laws of different variables (see below) were described by Overduin and Cooperstock (1998).

It must be remarked that the paper by Bertolami, considered the Brans-Dicke formalism, which is an alternative theory of gravity (see Chapter 10). Other similar treatments are due to Abdel-Rahman (1990), Özer and Taha (1986), and others. Special mention is to papers by Arbab I. Arbab (1997), and Beesham (1986).

The cosmological constant has been considered in the context of Quantum Field Theories, Kaluza-Klein penta-dimensional theories, quantum gravity, Grand-Unified Theories and supergravity. It is believed, from Gauge theories of elementary particles, that the cosmological constant is related to the mass of the Higgs boson. As a dynamical quantity, the cosmological term has been studied as anything that contributed to the energy density of the vacuum. Potential sources were identified for fluctuating vacuum energies, including scalar fields (Dolgov, 1983; Abbot, 1985; Peccei,Solà and Wetterich, 1987; Barr, 1987; Peebles and Ratra, 1988; Fujii and Nishioka, 1991; Frieman et al., 1995; Moffat, 1995); tensor fields (Hawking, 1984; Banks, 1984; Brown and Teitelboim, 1987; Dolgov, 1997); non-local effects (Linde, 1988; Banks, 1988); wormholes (Coleman, 1988); inflation phenomena (Tsamis and Woodart, 1995; Brandenberger and Zhinitsky, 1997); and also with cosmological perturbations (Abramo, Brandenberger and Mukhanov, 1997); etc.

For the benefit of the reader, we refer to Sahni and Starobinsky (2000), and Padmanabhan (2003).

The most popular solutions for a decaying Λ are picked in Table I of Overduin and Cooperstock (1998) definitive paper; Λ may be thought to vary not only with t^{-2} but with others powers of t ; in some other models it varies either with H (Hubble's parameter), q (deceleration parameter), or the scale factor, temperature, or in combinations of them, and, as we said above, in various kinds of power laws. [See next chapters].

8.2. Generalized Large Number Hypothesis

Berman (1992 b; 1994 b) extended the ideas due to the Nobel Prize winner Paul Adrien Maurice Dirac. When we divide Hubble's length, cH_0^{-1} by the Classical Electron's radius, $\frac{e^2}{m_e c^2}$, we find the adimensional number 10^{40} :

$$\frac{cH_0^{-1}}{\frac{e^2}{m_e c^2}} \sim 10^{40}. \tag{8.2.1}$$

This same order of magnitude appears when we divide the values for the electrostatic and the gravitational forces between an electron and a proton:

$$\frac{e^2}{Gm_p m_e} \sim 10^{40}. \tag{8.2.2}$$

On the other hand, its square, 10^{80} is roughly found for the total number of nucleons in the Universe, which we obtain by multiplying the observed mass density of the Universe by the cube of Hubble's length and then divide by the proton's mass:

$$\frac{\rho_0 \left(cH_0^{-1}\right)^3}{m_p} \sim (10^{40})^2 = 10^{80}. \tag{8.2.3}$$

Paul Dirac established his LNH (large numbers hypothesis) which says that those three numbers are large and not only coincidental: in fact, he said that the numbers are very large because the Universe is old, and they are time-varying and they are, in fact, time-varying, increasing with the age of the Universe so that, in the very early Universe these numbers were smaller. Lord Eddington (1935) found another large number 10^{40}, verbi gratia:

$$ch(m_p m_e/\Lambda_0)^{1/2} \sim 10^{40}, \tag{8.2.4}$$

where h stands for Planck's constant and Λ_0 is the maximum limit for the Λ constant in the present Universe. (Barrow, 1990).

Berman, substituted the number 10^{40}, by the function $N = N(t)$, while assuming the constant deceleration parameter law $H \propto t^{-1}$ and looked for the time variations of G, ρ and Λ. From (8.2.1) and the preceding time variation for H, we find:

$$N \propto t.$$

From (8.2.2) we obtain:

$$G \propto t^{-1}, \tag{8.2.5}$$

while from (8.2.3) we get:

$$\rho \propto t^{-2}, \tag{8.2.6}$$

and from (8.2.4) we find:

$$\Lambda \propto t^{-2}. \tag{8.2.7}$$

We have found again a solution with variable Λ as an elaboration of these ideas. More about this subject will be found in the next Chapters. It is interesting that, although the publishers did not claim for any damage, the GLNH papers by Berman, cited in the beginning of this Section, were in fact the same paper, but published twice, each time in a different journal, by mistake of this author. Someone accused Berman of having published, on purpose, for the second time, the same paper, and wrote his own paper on the same subject, dismissing Berman's calculations. In a reply to that accusation, I did show that his paper was inaccurate, because he first said that he was employing GRT, but then he confessed, in the sequel, that he was not.

8.3. Raychaudhuri's Equation and Λ- variable

From Einstein's equations for the Robertson-Walker's metric, it is easy to obtain an equation relating the sum $\rho + 3p$:

$$\tfrac{\kappa}{2}(\rho + 3p) = \Lambda - 3\tfrac{\ddot{R}}{R}. \tag{8.3.1}$$

If we impose, as usual in Cosmology, a perfect gas equation of state:

$$p = \alpha\rho, \tag{8.3.2}$$

with $\alpha =$ constant, and we take for granted formula (8.2.6) of last Section,

$$\rho = At^{-2}, \tag{8.3.3}$$

with A=constant, we obviously shall have,

$$p = \alpha A t^{-2}. \tag{8.3.4}$$

For constant deceleration parameter models we have:

$$\frac{\ddot{R}}{R} = \frac{(1-m)}{m^2} t^{-2}. \tag{8.3.5}$$

On combining all the above equations we find that:

$$\Lambda \propto t^{-2}. \tag{8.3.6}$$

The reader can see that all the above hypotheses are coherent with previous results, like those of the last Section's solution for variable Λ. Equation (8.3.1) is a particular one for the general Raychaudhuri's equation (Raychaudhuri, 1979).

8.4. Berman Model with Modified Energy Momentum Conservation

What follows, is contained in the paper by Berman (Berman, 1991b). We now consider Einstein's field equations, for a given Λ - term, which can be taken as the ρ equation (8.4.1) below, and the energy momentum conservation equation (8.4.3); the ρ equation is, as we know from Chapter 7 :

$$H^2 = \frac{8\pi G}{3}(\rho + \Lambda) - kR^{-2}. \tag{8.4.1}$$

The energy conservation equation arises by imposing $T^{\mu}_{\nu;\mu} = 0$, while we define the energy momentum tensor as a perfect fluid along with a Λ - term:

$$T_{\mu\nu} = pg_{\mu\nu} + (\rho + p)U_\mu U_\nu - \Lambda(t) g_{\mu\nu}. \tag{8.4.2}$$

Notice that we take the cosmological term as part of the matter contents. As long as the Einstein's tensor G^{μ}_{ν} obeys the conservation equation $G^{\mu}_{\nu;\nu} = 0$ due to Bianchi identities and the resulting energy tensor $\widetilde{T}_{\mu\nu} \equiv T_{\mu\nu} - \frac{\Lambda}{\kappa} g_{\mu\nu}$ is also conserved ($\widetilde{T}^{\mu}_{\nu;\mu} = 0$) we would find new field equations with $G^{\mu}_{\nu} = \kappa \widetilde{T}^{\mu}_{\nu}$, and no more questions asked, while $\Lambda = \Lambda(t)$.

The resulting conservation equation is:

$$\frac{d}{dt}(\rho R^3) + p\frac{dR^3}{dt} + R^3 \frac{d\Lambda}{dt} = 0. \tag{8.4.3}$$

We choose again the ideal gas equation of state:

$$p = \alpha\rho \ (\alpha = \text{constant}). \tag{8.4.4}$$

We further suppose that:

$$q = -\frac{\ddot{R}R}{\dot{R}^2} = const. = m - 1, \tag{8.4.5}$$

as in the constant deceleration parameter models, where also we have:

$$H = (mt)^{-1}, \tag{8.4.6}$$

and,

$$R = (mDt)^{1/m}. \tag{8.4.7}$$

We now try, as a solution, the laws for Λ and ρ variation found above:

$$\Lambda = \Lambda(t) = Bt^{-2} \qquad (B = \text{const}), \tag{8.4.8}$$

and,

$$\rho = \rho(t) = At^{-2} \qquad (A = \text{const }). \tag{8.4.9}$$

We shall see that, by plugging into the Einstein's equations it is found a coherent solution. Suppose that $k = 0$ (flat Universe); then:

$$A + B = \frac{3m^{-2}}{8\pi G}, \tag{8.4.10}$$

and

$$B = (1 + 3\alpha)\frac{A}{2}. \tag{8.4.11}$$

It seems that our individual results are consistent with each other.

If Λ was constant, our conservation equation above (8.4.3) would be like:

$$d(\rho V) + pdV = 0, \tag{8.4.12}$$

where V is any given volume containing a set of particles from the fluid, which we assume to be proportional to R^3. Equation (8.4.12) would stand for the entropy of such system, saying that:

$$TdS = 0,$$

where T is the absolute temperature and S is the entropy.

Now, in our case, equation (8.4.12) is not correct: instead, the entropy is not conserved but:

$$\frac{TdS}{dt} = -\gamma R^3 \frac{d\Lambda}{dt} = 2\gamma(mDt)^{3/m}Bt^{-3}, \tag{8.4.13}$$

where $V = \gamma R^3$ by hypothesis. (γ = constant). We see that the entropy is not constant and in fact, its increasing if $B > 0$:

$$\frac{dS}{dt} > 0.$$

An interesting case arises when $\alpha = -\frac{1}{3}$. This case corresponds, as it will be shown later, to a Machian solution for the Universe. This could, for instance, prevail in the very early Universe. In this case, we see from (8.4.11) that in our model $B = 0$, so that Λ is constant and the entropy is constant too. Remember that in equations (8.4.1) and (8.4.3) the solution Λ = constant is possible. Berman (2006) (2006 b) has shown that the Universe could grow out of nothing, and acquire the inflationary phase while solving Einstein's equation for $\alpha = -\frac{1}{3}$. In that reference, Berman made calculation with a constant Λ, arguing that this Λ was an average Λ practically constant in a very small period of time. Now, we see that Λ should be constant with this equation of state, in this model.

As the present Universe is thought as being in accelerated state we shall investigate now a possible solution with $q = -\frac{1}{2}$. Then,

$$m = q + 1 = \frac{1}{2}. \tag{8.4.12}$$

From the general $k = 0$ solution, we have:

$$A + B = \frac{3}{2\pi G}, \tag{8.4.13}$$

and

$$B = \left(\frac{1+3\alpha}{2}\right)A. \tag{8.4.14}$$

We now return to equations (8.4.1) and (8.4.3) from which we obtain:

$$\frac{\ddot{R}}{R} = -\frac{\kappa}{6}(\rho + 3p) + \frac{\Lambda}{3}. \tag{8.4.15}$$

A good picture for the above equation resides in defining, along with Whittaker (1935) and McCrea (1951), the density of gravitational mass:

$$\sigma = \rho + 3p,$$

so that the "total gravitational mass" is given by:

$$M \equiv \frac{4}{3}\pi R^3 \sigma, \tag{8.4.16}$$

and then we write for the acceleration of a particle on the "surface" of a sphere of radius R, and mass M, in a pseudo-Newtonian context,

$$\ddot{R} = -\frac{GM}{R^2} + \frac{\Lambda}{3}R, \tag{ 8.4.17}$$

where the acceleration has two parts, one due to Classical Gravitational Newtonian Theory, and the other is the acceleration due to the Λ – term.

We now calculate the deceleration parameter,

$$q = -\frac{\ddot{R}R}{\dot{R}^2} \equiv -H^{-2}\frac{\ddot{R}}{R} = H^{-2}\frac{GM}{R^3} - \frac{\Lambda}{3}H^{-2} = \frac{\Omega_m}{3} - \Omega_\Lambda, \tag{8.4.18}$$

where Ω_m and Ω_Λ stand respectively for the matter and the cosmological term density parameters.

The "flatness" condition, $\Omega_m + \Omega_\Lambda = 1$ means that the total contributions for the energy in the Universe is given by the $k = 0$ case in Friedmann-like models entails:

$$q = \frac{3}{2}\Omega_m - 1. \tag{8.4.19}$$

It seems that in terms of the supernova observations $\Omega_m < \frac{2}{3}$ so that $q \lesssim 0$.

We now return to the example given above, with $m = \frac{1}{2}$ and, consequently $q = -\frac{1}{2}$ which is typical of the possible acceleration of the Universe. The most cited values for the density parameters are:

$$\Omega_\Lambda \approx 0.7,$$

and

$$\Omega_m \approx 0.3.$$

We shall return to these numerical quantities, in next Chapter.

8.5. Abdel-Rahman–Arbab–Berman Model

The following text refers to Berman's paper (Berman, 1991).

In an alternative setting we have shown above that from GLNH (Generalized Large Number Hypothesis), Berman showed that one expects, if $\dot{\Lambda} \neq 0$, that we shall have $\dot{G} \neq 0$. To my recollection the first one to calculate the relationship between both variations while conserving the energy tensor, was A. Beesham though now this calculation is a triviality. The "trick" is to keep the strict General Relativistic energy tensor conservation equation, namely,

$$\dot{\rho} = -3H(\rho + p), \tag{8.5.1}$$

and then, we must cancel the following term:

$$\rho\dot{G} + \frac{\dot{\Lambda}}{8\pi} = 0. \tag{8.5.2}$$

Now Einstein's equations remain strictly valid:

$$3\ddot{R} = -4\pi G(t)R\left[3p + \rho - \frac{\Lambda(t)}{4\pi G(t)}\right], \tag{8.5.3}$$

and,

$$3\dot{R}^2 = 8\pi G(t)R^2\left[\rho + \frac{\Lambda(t)}{8\pi G(t)}\right] - 3k. \tag{8.5.4}$$

We again impose $q = m - 1 =$ constant, so that:

$$R = (mDt)^{1/m} \quad \text{with } m \neq 0 \text{ and } D = \text{constant}, \tag{8.5.5}$$

and

$$H = (mt)^{-1}. \tag{8.5.6}$$

Consider $k = 0$. It is easy to show that the following solution applies:

$$G\rho = At^{-2}, \qquad (A = \text{constant}) \tag{8.5.7}$$

$$\Lambda = Bt^{-2}, \qquad (B = \text{constant}) \tag{8.5.8}$$

From the field equations we check that:

$$8\pi A + B = 3m^{-2}; \tag{8.5.9}$$

$$G = Ft^{B/(4\pi A)}; \quad (F = \text{constant}) \tag{8.5.10}$$

and,

$$\rho = \tfrac{A}{F} t^{-(2+\frac{B}{4\pi A})}. \tag{8.5.11}$$

As we know from the present Universe, $F > 0$, so that $A > 0$ too. This makes $G > 0$ and $\rho > 0$ which is the so called weak energy condition.

From the conservation equation (8.5.1) we obtain:

$$p = \tfrac{A}{3F}\left[m(2 + \tfrac{B}{4\pi A}) - 1\right] t^{-(2+\frac{B}{4\pi A})}. \tag{8.5.12}$$

On comparing (8.5.11) with (8.5.12) we can see that we found a solution for a perfect gas equation of state of the type,

$$p = \alpha\rho, \tag{8.5.13}$$

with,

$$\alpha = \tfrac{1}{3}\left[m(2 + \tfrac{B}{4\pi A}) - 1\right]. \tag{8.5.14}$$

For an expanding Universe, we choose from (8.5.9),

$$m = \left[\tfrac{3}{B+8\pi A}\right]^{1/2} > 0. \tag{8.5.15}$$

The initial singularity, nevertheless, is a necessary feature of our show!

8.6. Chen and Wu's Model

Chen and Wu (1990) hinted that Λ could vary with a power of the scale factor. Özer and Taha (1986) had proposed a similar hypothesis with another justification. Though we agree with the general idea of Chen and Wu's paper, we take a different approach, which reaches the same result.

From dimensional analysis, we know that Λ is given by the inverse square of length. For Planck's time, we write:

$$\Lambda_{Pl} = \gamma L_{Pl}^{-2}. \quad (\gamma = \text{constant of order unity}) \tag{8.6.1}$$

For any other epoch, if the scale factor is predominant for the variation of Λ, then:

$$\Lambda = \gamma L_{Pl}^{-2} \left[\frac{L_{Pl}}{R}\right]^n. \tag{8.6.2}$$

Notice that we are imposing a power law kind of dependence. Notice also that, for $R = L_{Pl}$ we recover relation (8.6.1).

From the above, we find:

$$\Lambda = \gamma \frac{(L_{Pl})^{n-2}}{R^n}. \tag{8.6.3}$$

As our Physics is entirely Classical, we would not like to have a term that depended on Planck's constant, so that we impose $n-2=0$. We are thus led to consider as our variation law:

$$\Lambda = \gamma R^{-2}. \tag{8.6.4}$$

In order to show the richness of this approach, we shall make some calculations right now. Let us call

$$\alpha \equiv \frac{3}{8\pi G},$$

and

$$\Lambda = 3\gamma R^{-2}. \tag{8.6.5}$$

Now, suppose that the equation of state is that of a Machian Universe as proposed by Berman and Marinho Jr (2001) according to Chapter 9. Then,

$$k = 0 \quad \text{(flat Universe) and,} \quad p = -\tfrac{1}{3}\rho. \tag{8.6.6}$$

The two equations for Cosmology are:

$$H^2 = \alpha^{-1}\rho - \frac{\Lambda}{3}. \tag{8.6.7}$$

and, as in Section 8.4 above,

$$\frac{d}{dt}(\rho R^3) - \frac{1}{3}\rho \frac{dR^3}{dt} + \frac{\alpha}{3} R^3 \dot{\Lambda} = 0. \tag{8.6.8}$$

With the constant deceleration parameter model, we obtain, with:

$$R = R(t) = (mDt)^{1/m} \quad (m = \text{constant}). \tag{8.6.9}$$

$$H = \frac{1}{mt} = \frac{1}{(1+q)t}, \tag{8.6.10}$$

$$\rho = \rho_0 t^{-2/m}. \tag{8.6.11}$$

The following solution:

$$m = 1, \tag{8.6.12}$$

and,

$$\frac{8\pi G}{3}\rho_0 + \gamma D^{-2} = 1. \tag{8.6.13}$$

Now we find:

$$q = 0,$$

and

$$\Lambda = 3\gamma R^{-2} = \frac{3\gamma}{D^2} t^{-2}. \tag{8.6.14}$$

Suggested reading:

See the following papers, written by Brazilian Cosmologists, in the wake of my own papers: Maia and Silva(1994), Silveira and Waga(1994), Torres and Waga(1996), Silveira and Waga(1997), Carvalho, Lima and Waga(1992), Waga(1993), Salim and Waga(1993), Lima and Carvalho(1994), Lima and Maia(1994), Lima and Trodden(1996), Lima(1996). See also, the papers by John D. Barrow, with collaborators or alone, cited in the References at the end of this book.

Chapter 9

Energy of the Universe and the Λ - Problem

9.1. Energy of the Universe

The pioneer works of Nathan Rosen(1994), Cooperstock and Israelit(1995), showing that the energy of the Universe is zero, by means of calculations involving pseudotensors, and Killing vectors, respectively, are here given a more simple approach. We shall show that the energy of the Robertson-Walker's Universe is zero, that this Universe is Machian, and that it may lie inside a black-hole, thus hinting altogether to a solution of the cosmological constant problem.

Consider Minkowski's metric,

$$ds^2 = dt^2 - \left[dx^2 + dy^2 + dz^2\right]. \tag{9.1.1}$$

This is an empty Universe, except for test particles. We agree that its total energy is zero(Weinberg, 1972).

Now consider the expanding flat metric:

$$ds^2 = dt^2 - R^2(t)\left[dx^2 + dy^2 + dz^2\right]. \tag{9.1.2}$$

Here, $R(t)$ is the scale-factor. At any particular instant of time, $t = t_0$, we may define new variables, by the reparametrization,

$$dx'^2 \equiv R^2(t_0)\,dx^2, \tag{9.1.3}$$

$$dy'^2 \equiv R^2(t_0)\,dy^2, \tag{9.1.4}$$

$$dz'^2 \equiv R^2(t_0)\,dz^2, \tag{9.1.5}$$

$$dt'^2 \equiv dt^2.$$

Then,

$$ds^2 = dt'^2 - \left[dx'^2 + dy'^2 + dz'^2\right]. \quad (9.1.6)$$

The total energy of this "new" Universe, which is Minkowskian, is altogether, $E = 0$. We remember that in order to obtain the energy of a physical system, we have to consider a definite fixed instant of time. Thus, the flat Robertson-Walker's Universe reflects a zero-total energy.

9.2. The Energy of R. W's Universe

Consider now the metric:

$$ds^2 = dt^2 - \frac{R^2(t)}{\left[1+\frac{kr^2}{4}\right]^2}\left[dx^2 + dy^2 + dz^2\right]. \quad (9.2.1)$$

Here, $k = 0$ yields the flat case, already studied. When $k = \pm 1$, we have finite closed or infinite open Universes.

We want to calculate its energy. Consider $k = \pm 1$ (finite / open Universes). We are allowed to choose the way into making the calculation, so we choose a fixed value $\bar{r}$ of the radial coordinate, for which we reparametrize the metric:

$$dx'^{i2} \equiv \frac{R^2(t_0)dx^{i2}}{\left[1+k\frac{\bar{r}^2}{4}\right]^2} \quad (i = 1,2,3 \quad (9.2.2)$$

For this value of $r = \bar{r}$, the reparametrized metric has zero energy value, by the same token as in Section 9.1 above. Now we sum for all other values of r, obtaining an infinite sum of zeros, which yields a total energy, zero valued.

9.3. The Machian Universe

A more general treatment of the Machian Universe, then the one in this Chapter, will be presented in Section 10.6.

Weinberg(1972) has commented that, on experimental grounds, the known Universe obeys the Machian condition(Whitrow, 1946; Whitrow and Randall, 1951),

$$\frac{GM}{c^2R} \sim 1. \quad (9.3.1)$$

Here, M and R stand as the mass of the causally connected Universe, with radius ("scale-factor") of the Universe.

Adler et al.(1975) have calculated the energy of a spherical "ball", giving

$$E = Mc^2 - \frac{GM^2}{2R}. \quad (9.3.2)$$

As in the beginning of the Universe ($t = 0$) there was "nothing", we state that the Universe had zero energy. If energy is conserved for the Universe, it is reasonable to conjecture that it always will have zero energy.

Applying to the zero-energy of the Universe, we find

$$R = \frac{GM}{2c^2}. \tag{9.3.3}$$

If the "sphere" is representable by Schwarzschild's metric, we know that

$$R = \frac{GM}{2c^2} < \frac{2GM}{c^2} = R_{schw}. \tag{9.3.4}$$

In other words, the Universe, in our picture,

a) has zero-total energy;
b) lies inside a black-hole;
c) is Machian;
d) its radius is proportional to its mass.

We know from a paper by Berman and Marinho Jr(2001), and as we shall show in Section 9.7, that the primordial equation of state, for a Universe enjoying the Machian property, is $p = -\frac{1}{3}\rho$ where p. ρ stand for cosmic pressure and energy density.

Now, if Einstein's field equations are valid, for this equation of state, it is shown in Section 9.8 that with a non-null cosmological constant, the scale-factor R increases steadfastly, beginning from $R = 0$. According to (9.3.4), the mass M also increases, in the same proportion, from $M = 0$ at $t = 0$ when $R(0) = 0$, onwards. Even if $\wedge = 0$, $R(t)$ increases (linearly) with time (see Berman and Marinho, 2001). The instant $t = 0$ is the initial epoch of the Universe, in the purely Classical description. We must remember that within Planck's Universe, we can not ascribe definite values to the physical quantities involved. Instead, we consider, the Classical picture extended into Planck's Universe, as representing the average of the otherwise fluctuating physical values, when we go back in time.

It costs nothing to create the Universe: no expenditure either of mass or energy. In the next sections, we refine the arguments above.

9.4. Guessing a Solution Anew for the Λ - Problem

Our picture of the sphere, as representing the Universe, must be made more accurate, by including one more term into the total energy, which is the energy of the $\wedge$−term.

So we write:

$$E = Mc^2 - \frac{GM^2}{2R} + \frac{\wedge c^2 R^3}{6G} = 0\,. \tag{9.4.1}$$

where the $\wedge$ - term comes from multiplying the energy density of the vacuum, $\frac{\wedge}{K}$, by the volume of the sphere.

We now make a minor change in the dependence of the scale-factor to the mass, and we postulate now, for instance:

$$R = \frac{GM}{6c^2}. \tag{9.4.2}$$

If we suppose that

$$\wedge = \frac{72\,c^2}{R^2}, \tag{9.4.3}$$

then equation (9.4.1) is automatically obeyed, for we shall have

$$E = Mc^2 - 3Mc^2 + 2Mc^2 = 0.$$

Relation (9.4.3), namely, assuming that $\Lambda \propto R^{-2}$, may be a solution to the $\Lambda -$ Problem, i.e., of explaining why Λ was 10^{122} times larger during inflation than for the present age of the Universe, assuming, of course, that $R = R(t)$ is the "radius" of the Universe.

We refer to next sections, for more results for the scale factor and a positive lambda constant.

9.5. Comment on the Λ - Problem

In pages 90 and 91, Stephen Hawking(2001) describes inflation as an accelerated expansion of the Universe, immediately after the creation instant,while the Universe, as it expands,borrows energy from the gravitational field to create more matter. According to his description of inflation, the positive matter energy is exactly balanced by the negative gravitational energy, so that the total energy is zero,and that when the size of the Universe doubles, both the matter and gravitational energies also double, keeping the total energy zero (twice zero). **Our task will be to explain why the Universe obeys such characteristics, while being born and attaining an inflationary phase**. The Machian condition, represented by Whitrow-Randall-Brans-Dicke-Sciama relation will be employed,in order to find the equation of state in the very early Universe and afterwards, by means of Einstein´s field equations, we shall show that an exponential inflationary scenario is obtained in the Classical picture.

By postulating a Universe where GM/R would remain constant, and near unity, where M and R stand respectively for the mass and scale factor, Berman and Marinho Jr.(2001) have recently shown that the very early Universe might not have been radiation dominated, but could have attained a phase with equation of state such that cosmic pressure should equal minus 1/3 of the energy density.

In the same reference, a calculation has been done, with Einstein's field equations for General Relativity theory, and Robertson-Walkers metric, whereby the scale factor varied linearly with time, in the case of null cosmological constant.

The present author worked the positive cosmological constant case, and it is found that the same equation of state, if valid for the very early Universe, causes the scale factor to become exponentially increasing, thus pointing out to a graceful entrance into inflationary phase.

The cosmological constant is associated with a crucial problem in particle physics and cosmology. It was introduced by Einstein in General Relativity in order to obtain a stationary solution to the Universe. However, as it became clear experimentally that the Universe is expanding, the cosmological constant term in the Einstein's equations would not be necessary, in principle. In addition, a stationary Universe as a solution of the Einstein field equations can be obtained even without the cosmological constant term (de Sitter Universe). We must recognize that the de Sitter Universe is static in the sense that the energy density and curvature are constant, but the necessary cosmological constant term causes test particles to accelerate away one from another.

On the other hand, it was shown for the first time by Zeldovich in the year 1967, by means of a series of interesting papers(Zeldovich, 1981), that the presence of the cosmological constant in Einstein's equations is correct, but this term has another interpretation: it is associated to the vacuum energy density. When we take the inverse ratio between the observed value of the cosmological constant and the value predicted by gauge theories of particle physics we find the amusing discrepancy of $\sim 10^{122}$. It is the cosmological constant problem (see reviews in Weinberg, 1989; Padmanabhan, 2002; Carroll, 2001). There is, at present, no satisfactory theory which points to the solution of this problem. As explained below, in this report, we treat the problem briefly,within a realistic perspective.

In the last years, many attempts has been spent in order to advance in this question. Recent astronomical observations showed that the cosmological constant is positive, which implies that the expansion of the Universe is accelerated (Perlmutter, 1999; Riess et al., 1998; Falco et al., 1998; Bernardis et al.. 2000; Netterfield et al., 2002; Halverson et al., 2002; Lee et al., 2001). The vacuum energy density, i. e., the cosmological constant, affects directly the structure and evolution of the Universe. Therefore, advances in this question can lead to the solution of several problems in the intersection between Particle Physics and Cosmology.

The equation of state shall be dealt with in Section 9.7, and it points out to a cosmic strings phase (see Berman and Marinho, 2001).

In the following sections, we shall discuss the Brans-Dicke(1961) relation, also derived by Whitrow(1946), and Whitrow and Randall(1951), the connection of this relation with the zero-total energy conjecture for the whole Universe, made 40 years ago by Feynman(1962-3), and the particular form devised by Hawking in his new best-selling book(2001). Then we derive the aforementioned equation of state for the v.e.U. (very early Universe), and we solve Einstein's field equations for a perfect fluid Robertson-Walker's metric, in the presence of a positive cosmological constant. We shall show that, the zero-total energy conjecture, may include a lambda term, which is a time-decreasing function of the scale-factor, and thus it might be solving the cosmological constant problem. This problem consists in the puzzling discrepancy, of more than a hundred orders of magnitude, between its present possible value, and the value required, for the v.e.U., in the current Particle Physics calculations. The reported derivation is true only if theorists can provide a model to explain why lambda varies like this, why the resulting "constant " has the value it does today, and how the formation and evolution of galaxies can be accommodated. For simplicity, the model to be outlined in section III and the remainder of this paper, actually results in a non-varying cosmological constant. If the reader wanted to apply the correct formalism, use could be made of the set of equations developed by Berman(1991).

9.6. Brans-Dicke Relation

Brans-Dicke relation(Brans and Dicke, 1961; Whitrow, 1946; Whitrow and Randall, 1951), has been widely discussed in the literature(see, for instance, Weinberg's book(1972)):

$$GM/R \approx 1 \tag{9.6.1}$$

(we employ relativistic units $c = 1$). The above relation is justified by plugging the estimated values for the radius and mass of the Universe, at present. We can obtain this relation by considering that the total energy of a sphere is given by adding the energy of its matter content, M, plus the negative gravitational self energy, which is given in Newtonian Physics by $-3GM^2/(5R)$.

We drop the factor $3/5$ from the resulting equation, because this is a back of the envelope estimate, for the Universe. Adler, Bazin and Schiffer(1975) made a general relativistic calculation for such a sphere, obtaining a factor $\frac{1}{2}$ instead of $3/5$, so we are safe when we write for the total energy of the Universe, E, approximately,

$$E \approx M - GM^2/R.$$

The zero total energy is thus a consequence of Brans-Dicke expression for Machian property. It should be remarked that, when dealing with such energy calculations, for the Universe, one must remember that the calculation done in Weinberg(1972) was on the Schwarzschild's metric in a flat background space.

It is obvious that Hawking thought on this formula, and that $R = \varepsilon GM$ where $\varepsilon =$ constant, when he said that doubling the size of the Universe, the matter energy and the potential energy, plus the total energy-all double, but twice zero is still zero.

As later on we shall introduce Einstein's equations with a cosmological constant, let us briefly take into account the energy density of the vacuum, $\frac{\Lambda}{\kappa}$, where $\kappa = 8\pi G$ (Einstein's gravitational constant).

The energy of such sphere should now be:

$$E = M - GM^2\ /(2R) + \Lambda R^3\ /(6G). \tag{9.6.2}$$

We have used Adler et al. result, augmented by the lambda energy. The latter is a positive contribution, for a positive lambda constant. Hawking's statements will stand correct if we add the hypothesis that lambda decays with the inverse square of the scale-factor. This could perhaps solve the cosmological constant problem, as we mentioned in our introduction. We refer to the paper by Chen and Wu(1990) for arguments in favour of this dependence, and why the resulting "constant" has the value it does today (see also Özer and Taha´s papers(1986)(1987)).

Hawking must have also envisaged a Universe obeying the relations

$$R = \varepsilon GM, \tag{9.6.3}$$

$$\Lambda = 3\varepsilon^{-1}\ [\varepsilon^{-1} - 2]R^{-2}. \tag{9.6.4}$$

where $\varepsilon < \frac{1}{2}$ for a positive lambda, or else we could not have the "doubling" property mentioned by him along with the above formulae for the total energy. Now, this may mean that Hawking's Universe lies inside its Schwarzschild's radius, or, as he says, in a nutshell-is that a black hole???.

9.7. Equation of State for the Very Early Universe

Fujii and Maeda(2003) examine briefly the principles of inflationary theory (Guth, 1981). In the v.e.U., a rapid expansion must have taken place, driven by the vacuum energy, or by a scalar field falling down a potential slope.

We shall show that the cosmological constant does all the job, along with Einstein's field equations, for a Robertson-Walker's metric, provided that the equation of state derived by Berman and Marinho Jr. be accepted as valid in the v.e.U.

On dimensional grounds, by combining the constants for the speed of light in vacuum, Newton's gravitational, and Planck's, we find the length and mass of the primordial Universe, named after Planck; and Berman(1994) has remarked that they obey the exact Machian relation,

$$GM = R. \tag{9.7.1}$$

Extending this relation to the classical domain, Berman and Marinho(2001) applied the balance of energy equation,

$$p = -dM/dV, \tag{9.7.2}$$

in order to show that the relation between energy density and cosmic pressure, was

$$p = -\rho/3. \tag{9.7.3}$$

Due to the minus sign, this is not a radiation dominated Universe, as was presumed by many authors. It rather represents a cosmic strings phase (Kolb and Turner, 1990).

The derivation of the equation of state runs as follows:

$$R = GM.$$

The volume of a 3-dimensional space is given by:

$$V = \alpha R^3, \text{ with } \alpha \text{ a convenient constant.} \tag{9.7.4}$$

It should be remarked that, for open and flat Universes, one must apply the Brans-Dicke relation,to the finite mass-energy contained in a causal patch of the Universe within a radius R — or else, we must face an infinite volume.

Then,

$$\rho = M/V = (\alpha G R^2)^{-1}. \tag{9.7.5}$$

From the balance of energy equation, and the above relations,

$$p = -dM/dV = (3\alpha R^2 G)^{-1}. \tag{9.7.6}$$

On comparing the obtained values for cosmic pressure and energy density, the equation of state is:

$$p = -\rho/3.$$

Please pay attention that we have not entered with Einstein's field equations up to now.

The same equation of state is obtainable by taking the Universe to obey,as in last section, the relation $GM/R = 1/\varepsilon$, where ε is a constant.

9.8. Creation of the Universe out of Nothing

Let us consider Robertson-Walker's metric:

$$ds^2 = dt^2 - \frac{R^2}{[1+kr^2/4]^2}[dr^2 + r^2 d\theta^2 + r^2 \sin^2\theta d\phi^2] \,, \tag{9.8.1}$$

where k represents the tricurvature $(0, +1, -1)$.

We can check that, Einstein's field equations read, for the above metric, and a perfect fluid energy-tensor:

$$\kappa\rho = 3H^2 + 3kR^{-2} + \Lambda, \tag{9.8.2}$$

$$\kappa p = -2R^{-1}\ddot{R} - H^2 - kR^{-2} - \Lambda, \tag{9.8.3}$$

where $\rho, \kappa, \Lambda, p, and\ H$, stand respectively for the energy density, Einstein's gravitational constant, cosmological constant, cosmic pressure, and Hubble's parameter $(H = \dot{R}/R)$.

On solving (9.8.2) and (9.8.3) for the derived equation of state, we find:

$$R(t) = exp(\ \beta t\ ) - exp(-\beta t), \tag{9.8.4}$$

where

$$\beta = \tfrac{1}{2}\sqrt{\left(\tfrac{8\pi}{3\alpha} - k\right)}, \tag{9.8.5}$$

and

$$\Lambda = 3\beta^2. \tag{9.8.6}$$

As we are in face of rapid expansion, the negative exponential can be neglected from our solution for the scale-factor. This solution shows a graceful entrance into inflationary epoch (Guth, 1981).

Based on the resulting solution that makes $R(0) = 0$ at $t = 0$, and which does not avoid the problems associated with infinite energy densities at the initial instant of the Universe, we must recognize that Einstein's field equations do NOT apply when Quantum phenomena come to play a vital role. The initial singularity, is just a property of the mathematical nature in the Classical approach.

There is no much harm in the use of a constant lambda, which we may regard to represent an average value, valid in the "small" time interval of inflationary phase. Berman(1991) worked the full field equations for lambda variable, for a general perfect gas equation of state, and a perfect fluid energy-tensor.

As it is well-known, for a finite RW´s metric,

$$\alpha = 2\pi^2.$$

This makes our solution oscillatory, and with negative lambda. We would not have an exponential, but the sinusoidal period could be very large, so we can keep only, as a solution, the positive slope part of the curve, and still have a model for the v.eU. However, adjusting this solution would require a fine tuning that renders our $k = +1$ case rather implausible.

Furthermore, we would need to adjust different constant coefficients multiplying the positive and negative exponentials in the solution for the scale factor. Thus, we could avoid the initial singularity at $t = 0$, so that $R(t)$ would have its minimum value of the order of Planck´s length. We prefer NOT to do this, for we understand that the initial singularity, in the Classical domain, is a necessary and typical consequence, caused indeed by the fact that Einstein´s theory should be substituted by Quantum Gravity theories in dealing with Planck´s Universe.

Because β is then a real-valued quantity, flat and open models are free to show exponential inflation, and pose no additional problems (Berman and Trevisan, 2001).

We have, so far, reviewed the derivation of the equation of state that might apply for the very early Universe, and have shown that it leads towards a graceful entrance into inflationary epoch, all these facts depending just on the Machian relation for the Universe, first derived by Whitrow, and Whitrow and Randall, and then popularized by Brans and Dicke. Dennis Sciama also arrived to this relation, before Brans and Dicke, in a different gravitational theory (Sciama, 1953).

We have interluded our report with comments on a paragraph in Prof. Hawking's new book(2001), where he gives possible details on the energy of the inflationary Universe and its physical constituent terms (matter energy, and potential gravitational energy). We added a lambda energy term to his picture of a zero total energy for the Universe. It is clear that Hawking presupposes a Machian Universe altogether.

It should be clear that Brans-Dicke relation must be taken in the modified form, i.e., $R = \varepsilon GM$. Remember that ε is of order unity.

We point out, that the cosmological calculations in this paper revert into an oscillatory model for $k = +1$, with accommodation of a negative lambda into our framework of the modified Brans-Dicke relation ($\varepsilon > \frac{1}{2}$). Thus, we may have a black-hole anyway. Further discussion of this case, and also of the open Universe ($k = -1$), may become necessary, and we postpone it to a future paper. However, the graceful <u>exit</u> from inflationary phase is a solved topic(Kolb and Turner, 1990).

In our last model, we find that the scale factor increases from its zero value when $t = 0$ and $M = 0$ while the total energy input was also zero. This is creation out of nothing. Nevertheless, we presuppose that Einstein's equations are valid along with the Robertson-Walker's metric for the Universe since $t = 0$ onwards. In the Quantum Universe, of course, Planck's constant h would have a specific role.

It must be remarked that a non-null lambda constant value, and also of dark energy, in order to get a negative value for the cosmic pressure, points to an accelerating present Universe. The origin of the lambda concept should belong to the Quantum Gravity chapter of particle physics in the very early Universe.

Another comment is necessary: from the R^{-2} – dependence of the several energy densities considered as relevant for the Machian Universe, we may state that, by varying alike, they keep the same relative amounts of each type of energy during the entirely life of the Machian Universe; this means that, given a present Λ – driven Universe, it would be equally driven at earlier epochs.

9.9. Cosmological Parameters

Following Weinberg (1972) we write the cosmological equations with a Λ-term, by making the following substitutions into the $\Lambda = 0$ Einstein's equations:

$$\rho \rightarrow \rho + \tfrac{\Lambda}{\kappa}, \tag{9.9.1}$$

and,

$$p \rightarrow p - \tfrac{\Lambda}{\kappa}. \tag{9.9.2}$$

We choose Einstein's equation for the energy density, plus the energy momentum conservation equation which are equivalent to both Einstein's equations, one for ρ, the other for p :

$$\dot{R}^2 = \tfrac{\kappa}{3}\rho R^2 + \tfrac{\Lambda}{3}R^2 - k, \tag{9.9.3}$$

and,

$$\tfrac{d}{dR}\left[\left(\rho + \tfrac{\Lambda}{\kappa}\right) R^3\right] = -3R^2\left(p - \tfrac{\Lambda}{\kappa}\right). \tag{9.9.4}$$

We impose a perfect gas equation of state, $p = \alpha\,\rho$ (α = constant). Then,

$$\tfrac{d}{dR}\left(\rho R^{3(1+\alpha)}\right) = \tfrac{1}{\kappa}R^{3(1+\alpha)}\tfrac{d\Lambda}{dR}\ ; \tag{9.9.5}$$

and,

$$\ddot{R} = -\tfrac{\kappa}{3}\left(\tfrac{1+3\alpha}{2}\right)\rho R + \tfrac{\Lambda}{3}R. \tag{9.9.6}$$

We now find the deceleration parameter,

$$q = \tfrac{1}{2}(1 + 3\alpha)\Omega_m - \Omega_\Lambda, \tag{9.9.7}$$

where,

$$\Omega_m \equiv \tfrac{\kappa\rho}{3H^2} \equiv \tfrac{\rho}{\rho_{crit}}\ ; \tag{9.9.8}$$

$$\Omega_\Lambda \equiv \tfrac{\Lambda}{3H^2}\ ; \tag{9.9.9}$$

$$\rho_{crit} \equiv \tfrac{3H^2}{\kappa}. \tag{9.9.10}$$

The above definitions for Ω_m and Ω_Λ are the components of energy density for matter and the cosmological constant, divided by the critical density ρ_{crit} which is the value of the energy density which corresponds to its value in a flat $(k = 0)$, pressureless $(p = 0)$, Friedmann-Robertson-Walker model $(\Lambda = 0)$.

From equation (9.9.3) and the definitions (9.9.8) (9.9.9) and (9.9.10) we find:

$$k = \dot{R}^2[\Omega_m + \Omega_\Lambda - 1]. \tag{9.9.11}$$

The flat case, which is preferred from inflationary theory is given by$k = 0$or:

$$\Omega_m + \Omega_\Lambda = 1. \tag{9.9.12}$$

Supernovae observations lead to a value $\Omega_\Lambda \approx 0.7$; if the Universe is flat $\Omega_m \approx 0.3$. Such a value for Ω_m seems a little high, so that cosmologists are looking for missing mass in the form of dark matter; the obvious candidates are massive neutrinos, though the subject is open to other conjectures (axions, light-supersymmetric particle, etc). As we remarked earlier, the value for Ω_Λ quoted above, fits the cosmologists need to get a dark energy value that makes cosmic pressure negative, thus pointing to an accelerating present Universe.

If, for the Universe $p \cong 0$, which means that $\alpha \cong 0$, the deceleration parameter must be, according to (9.9.7):

$$q \cong 0.15 - 0.7 \cong -0.55.$$

With Berman models (q =constant) we must have $m = q + 1 \geq 0$, so that it corresponds to an expanding Universe, so that, $q \geq -1$, and thus we expect, from the latest observations,

$$0 \geq q \geq -1.$$

In the general case of a non-flat curved Universe $(k \neq 0)$ we may define a "curvature" density parameter $\Omega_k \equiv -\frac{k}{\dot{R}^2}$, and write, formally:

$$\Omega_m + \Omega_\Lambda + \Omega_k = 1. \tag{9.9.13}$$

In the above form, the matter and lambda density parameters must be taken care while remembering that the curvature of the Universe may also play a definite role.

9.10. Gravitational Density with Λ - Term

Besides being a new form of vacuum energy, the Λ-term is interpreted as a repulsive field of forces in the Universe. We now give a specific meaning to this interpretation.

From the Cosmological equations above, one finds, for the Robertson-Walker's metric, the equation:

$$\ddot{R} = -\frac{8\pi G}{3}\left[\frac{\rho + 3p}{2}\right]R + \frac{\Lambda}{3}R. \tag{9.10.1}$$

We define the total gravitational mass density σ, as:

$$\sigma = \rho + 3p. \tag{9.10.2}$$

If this interpretation is correct the total gravitational mass M should be given by:

$$M \equiv \tfrac{4}{3}\pi R^3 \sigma = \tfrac{4}{3}\pi R^3 (\rho + 3p).$$

Then,

$$\ddot{R} = -\tfrac{GM}{R^2} + \tfrac{\Lambda}{3} R = a_G + a_\Lambda, \tag{9.10.3}$$

where this "would be" acceleration of a particle on the "surface" of a sphere of radius R and mass M, is subject to the attractive Newton's gravitation law acceleration a_G, and the repulsive Λ – field of accelerations a_Λ. In other words, the Λ–force is responsible for a repulsive acceleration:

$$a_\Lambda = \tfrac{\Lambda}{3} R. \tag{9.10.4}$$

The above definition of σ, and M, take in consideration that pressure carries weight.

Whittaker (1935), and later McCrea (1951), have included the cosmological constant as a negative mass density, redefining the density of gravitational mass as:

$$\sigma_W = \rho + 3p - \tfrac{2\Lambda}{\kappa}. \tag{9.10.5}$$

With the above definition, and Einstein's field equations, we find for the deceleration parameter the expression:

$$q = \tfrac{4\pi}{3} G H^{-2} \sigma_W. \tag{9.10.6}$$

From the above formula we can check that negative values for the deceleration parameter must obey the condition

$$\sigma_W < 0. \tag{9.10.7}$$

Gomide (1980) (1985) and Gomide and Uehara (1981) have pointed out that when,

$$\sigma q^{-1} \approx \rho, \tag{9.10.8}$$

we obtain Whitrow-Randall's relation, (see Whitrow, 1946; Whitrow and Randall, 1951):

$$\tfrac{4\pi}{3} G \rho H^{-2} \approx 1. \tag{9.10.9}$$

In fact, when we consider the mass M of a "large" sphere of radius R and we equate its inertial energy contents Mc^2 with the self gravitational energy $\tfrac{3}{5} G \tfrac{M^2}{R}$ we find, approximately:

$$\tfrac{GM}{c^2 R} \approx 1, \tag{9.10.10}$$

which means that inertia is equivalent to gravitation, in the form of Mach's principle: this is Brans-Dicke relation for the Machian Universe. Now we consider distances of the order of Hubble's length,

$$r_H \approx cH^{-1}, \tag{9.10.11}$$

and we write the cosmic density expression:

$$\rho \approx M \left[\tfrac{4\pi}{3} r_H^3\right]^{-1}, \tag{9.10.12}$$

we find the alternative Whitrow-Randall expression:

$$\tfrac{4\pi}{3} G\rho H^{-2} \approx 1. \tag{9.10.9}$$

Notice that with Gomide-Uehara approximation, (9.10.8), we obtained Whitrow-Randall relation, and indirectly Brans-Dicke relation, from Einstein's field equations, with $\Lambda \neq 0$.

A different approach can be followed from the exact GRT relation (9.10.6). If the driving factor in the Universe is the cosmological constant, we may discard energy density and pressure, in the calculation of σ_W :

$$\sigma_W = \rho + 3p - \tfrac{2\Lambda}{\kappa} \cong -\tfrac{2\Lambda}{\kappa}. \tag{9.10.10}$$

When we consider constant, or nearly constant deceleration parameters, we may use Berman's relation, now widely used in the literature:

$$H = (mt)^{-1} = [(1+q)t]^{-1}. \tag{9.10.11}$$

Relation (9.10.6) becomes now:

$$\Lambda \cong -3q\,(1+q)^{-2} t^{-2}. \tag{9.10.12}$$

As the present observations favor a positive Λ, the above formula needs a negative q. If Λ is constant, q would vary with time, in an unprecedented way; we conclude, then, that the only solution is adopting a time-varying Λ, and keeping a constant q.

Chapter 10

Miscellaneous Cosmological Models

10.1. Introducing Variational Principles

The following digression is not intended as a full treatment of the variational methods in Physics, but it should ease understanding of next Section. We follow closely the approach by Raychaudhuri et al (1992), which is, along with Raychaudhuri (1979), one of the best textbooks in the Theory of Gravitation, and Cosmology. Our purpose will be paving the way for the next Section.

Hamiltonian Mechanics is based on a scalar function $L = L\ (q_j, \dot{q}_j, t)$ where q_j are generalized coordinates and $\dot{q}_j$ the generalized velocities. The action integral A is defined as:

$$A = \int_{t_1}^{t_2} L\, dt, \tag{10.1.1}$$

where the integral is calculated between an initial and a final state. We find the actual motion of a particle, by infinitesimally varying the paths, until one finds the specific path for which A is stationary:

$$\delta A = 0. \tag{10.1.2}$$

One then finds Lagrangian equations.

For fields, we repeat the procedure, like follows. Consider a field described by independent variables ϕ^A, with $A = 1, 2, ...$:

$$\phi^A = \phi^A(x^\alpha).$$

We define:

$$\delta A \equiv \delta \int L(\phi^A, \phi^A_{,\alpha}) \sqrt{-g} d^4x = 0, \tag{10.1.3}$$

where the integrand is an invariant scalar, because L is a scalar density. It is understood that the boundary of the integration domain is a three-dimensional hypersurface, such that $\delta\phi^A = 0$ in the boundary, in analogy with the Classical Mechanics case.

This leads to:

$$\int \sum_A \left[\frac{\partial L}{\partial \phi^A}\delta\phi^A + \frac{\partial L}{\partial \phi^A_{,\alpha}}\delta\phi^A_{,\alpha}\right]\sqrt{-g}d^4x = 0.$$

There follows that:

$$\int \sum_A \left[\frac{\partial L}{\partial \phi^A}\sqrt{-g} - \frac{\partial}{\partial x^\alpha}\left(\frac{\partial L}{\partial \phi^A_{,\alpha}}\sqrt{-g}\right)\right]\delta\phi^A d^4x = 0. \qquad (10.1.4)$$

The integrand must be null, because the $\delta\phi^A$ are arbitrary, so that:

$$\frac{\partial L}{\partial \phi^A} - \frac{1}{\sqrt{-g}}\frac{\partial}{\partial x^\alpha}\left(\frac{\partial L}{\partial \phi^A_{,\alpha}}\sqrt{-g}\right) = 0. \qquad (10.1.5)$$

These are the equivalent Lagrangian equations in our field theory.

In the next Section, we shall need to consider a massless scalar field. For generality we shall derive now the Klein-Gordon equation for a massive scalar field; the obvious scalar Lagrangian, where only the first derivatives of the scalar field are considered, is given by:

$$L = \phi_{,\delta}\phi'^{\delta} + m^2\phi, \qquad (10.1.6)$$

where m is the mass term. The Lagrangian equations are:

$$m^2\phi - \frac{1}{\sqrt{-g}}\frac{\partial}{\partial x^\alpha}\left(\phi_{,\delta}g^{\delta\alpha}\sqrt{-g}\right) = 0. \qquad (10.1.7)$$

We now introduce the d'Alembertian, $\Box^2$, and find:

$$m^2\phi - \phi_{;\alpha}^{\alpha} \equiv m^2\phi - \Box^2\phi = 0, \qquad (10.1.8)$$

which is the desired equation.

Variational Principle for the matter fields

We now turn to the energy-momentum tensor, which describes the "matter" fields (via the energy density and pressure, for instance). Instead of variations of the fields variables, we consider variations of the metric tensor, in order to find a $T_{\mu\nu}$ whose covariant divergence vanishes:

$$T^{\mu\nu}_{;\nu} = 0.$$

This tensor will be identified with the energy-momentum tensor. By means of an infinitesimal coordinate transformation $x'^\alpha \to x^\alpha - \xi^\alpha$, we have $g_{\mu\nu} \to g'_{\mu\nu}$; we find explicitly:

$$g'_{\mu\nu} = g_{\alpha\beta}\frac{\partial x^\alpha}{\partial x'^\mu}\frac{\partial x^\beta}{\partial x'^\nu} = g_{\alpha\beta}\left[\delta^\alpha_\mu + \frac{\partial \xi^\alpha}{\partial x'^\mu}\right]\left[\delta^\beta_\nu + \frac{\partial \xi^\beta}{\partial x'^\nu}\right] \cong g_{\mu\nu} + \xi_{\nu;\mu} + \xi_{\mu;\nu} - \xi^\alpha\frac{\partial g_{\mu\nu}}{\partial x^\alpha} \cong$$

$$\cong g_{\mu\nu}(x'^\alpha) + \xi_{\nu;\mu} + \xi_{\mu;\nu}.$$

The reader should remember Section 5.4 where we defined the Lie derivative. Here we have:

$$\delta g_{\mu\nu} \equiv g'_{\mu\nu} - g_{\mu\nu}(x'^{\alpha}) = \xi_{\nu;\mu} + \xi_{\mu;\nu} = L_{\vec{\xi}}\, g_{\mu\nu}. \tag{10.1.9}$$

The last term is the Lie derivative of the metric tensor in the $\vec{\xi}$ direction.

We write the total action into two pieces, one, δA_1, containing the $\delta\phi^A$ variation; the other containing the $\delta g_{\mu\nu}$ variation. From the preceding theory, $\delta A_1 = 0$, so that:

$$\delta A = \delta\int L\sqrt{-g}d^4x = \int\left[\frac{\partial(L\sqrt{-g})}{\partial g^{\mu\nu}}\delta g^{\mu\nu} + \frac{\partial(L\sqrt{-g})}{\partial g^{\mu\nu}_{,\alpha}}\delta(g^{\mu\nu}_{,\alpha})\right]d^4x = 0. \tag{10.1.10}$$

Due to the product rule for derivatives,

$$\frac{\partial}{\partial x^{\alpha}}\left[\frac{\partial(L\sqrt{-g})}{\partial g^{\mu\nu}_{,\alpha}}\delta g^{\mu\nu}\right] \equiv \frac{\partial}{\partial x^{\alpha}}\left[\frac{\partial(L\sqrt{-g})}{\partial g^{\mu\nu}_{,\alpha}}\right]\delta g^{\mu\nu} + \frac{\partial(L\sqrt{-g})}{\partial g^{\mu\nu}_{,\alpha}}\delta(g^{\mu\nu}_{,\alpha}). \tag{10.1.11}$$

From the above two expressions we find:

$$\delta A = \int\left\{\left[\frac{\partial(L\sqrt{-g})}{\partial g^{\mu\nu}}\right]\delta g^{\mu\nu} - \frac{\partial}{\partial x^{\alpha}}\left[\frac{\partial(L\sqrt{-g})}{\partial g^{\mu\nu}_{,\alpha}}\right]\delta g^{\mu\nu}\right\}d^4x. \tag{10.1.12}$$

In the above expression, because of the Gauss theorem, we have neglected the total divergence of $\frac{\partial(L\sqrt{-g})}{\partial g^{\mu\nu}_{,\alpha}}\delta g^{\mu\nu}$, which, by volume integration, is transformed into a surface integral on the boundary of the domain, where $\delta g^{\mu\nu} = 0$.

We are now in position of defining a symmetric energy-momentum tensor $T_{\mu\nu}$, by means of:

$$\frac{1}{2}\sqrt{-g}T_{\mu\nu} \equiv \frac{\partial(L\sqrt{-g})}{\partial g^{\mu\nu}} - \frac{\partial}{\partial x^{\alpha}}\left[\frac{\partial(L\sqrt{-g})}{\partial g^{\mu\nu}_{,\alpha}}\right]. \tag{10.1.13}$$

Then,

$$\delta A = \frac{1}{2}\int T_{\mu\nu}\delta g^{\mu\nu}\sqrt{-g}d^4x = -\frac{1}{2}\int T^{\mu\nu}\delta g_{\mu\nu}\sqrt{-g}d^4x, \tag{10.1.14}$$

where we know that the δ operation is not affected by the metric tensor:

$$\delta g^{\mu\nu} = -g^{\mu\alpha}g^{\nu\beta}\delta g_{\alpha\beta}. \tag{10.1.15}$$

From (10.1.9) and (10.1.14) we find, due to the symmetry of the energy-momentum tensor:

$$\delta A = \int T^{\mu\nu}\xi_{\mu;\nu}\sqrt{-g}d^4x = -\int(T^{\mu\nu}\xi_{\mu})_{;\nu}\sqrt{-g}d^4x + \int(T^{\mu\nu}_{;\nu}\xi_{\mu}\sqrt{-g}d^4x), \tag{10.1.16}$$

where we used the Leibniz rule for the covariant derivative of a product.

The volume integral of the covariant divergence of $T^{\mu\nu}\xi_{\mu}$ can be converted into a surface integral at the boundary, where $\xi_{\mu} = 0$, so that $\delta g^{\mu\nu}$ also vanishes. As, in principle, inside the volume integral, the ξ_{μ} 's are arbitrary quantities, in order to have $\delta A = 0$ we need to impose:

$$T^{\mu\nu}_{;\nu} = 0.$$

As we already discussed in Section 3.2, because $T^{\mu\nu}$ is known to be conserved in Special Relativity (Minkowski's spacetime), we identify, as required, $T^{\mu\nu}$ as the energy-momentum tensor in General Relativity, without any inconsistency.

If L does not depend explicitly on $g^{\mu\nu}_{,\alpha}$, we return to (10.1.13) in order to find:

$$\tfrac{1}{2}\sqrt{-g}T_{\mu\nu} = \tfrac{\partial L}{\partial g^{\mu\nu}}\sqrt{-g} + L\tfrac{\partial(L\sqrt{-g})}{\partial g^{\mu\nu}}, \quad \text{or,}$$

$$T_{\mu\nu} = 2\tfrac{\partial L}{\partial g_{\mu\nu}} - Lg_{\mu\nu}\,. \tag{10.1.17}$$

So that, given L, we find directly $T_{\mu\nu}$, if the Lagrangian yields the "correct" field equations.

From the Lagrangian for the massive scalar field,

$$L = m^2\phi + \phi_{,\alpha}\phi^{,\alpha}, \text{ we find,}$$

$$T_{\mu\nu} = 2\tfrac{\partial L}{\partial g_{\mu\nu}} - Lg_{\mu\nu} = 2(\phi_\mu\phi_\nu - \tfrac{1}{2}g_{\mu\nu}\phi_{,\alpha}\phi^{,\alpha}) - m^2\phi^2 g_{\mu\nu}.$$

In the Brans-Dicke theory, we take $m = 0$ for the massless scalar field. (See next Section).

Variational Principle for the Gravitational Field

Now we turn our prescriptions in order to find a Gravitational variational principle with an action A_g, so that our previous action for matter fields, A_m will yield:

$$\delta A_g + \delta A_m = \delta(A_g + A_m) = 0, \tag{10.1.18}$$

where,

$$A_m = \int L_m\sqrt{-g}d^4x \quad ;\text{ and}$$

$$\delta A_m = \tfrac{1}{2}\int (T_{\mu\nu})_m \delta g^{\mu\nu}\sqrt{-g}d^4x. \tag{10.1.19 a}$$

For the gravitational field, we must find a Lagrangian so that we construct scalars from the metric tensor and its first and second derivatives, which are the only ones that arise into the Einstein tensor. From General Relativity Theory, we know one such scalar: R – the Ricci scalar. Although the Lagrangian will depend also on the second derivatives of the $g_{\mu\nu}$ these terms disappear on integration, by being transformed into surface integrals in the boundary of the domain, where the variation of the metric tensor goes to zero ($\delta g_{\mu\nu} = 0$). On writing:

$$L \equiv R \equiv R^i_i \equiv g^{ij}R_{ij}, \tag{2.4.5}$$

we have:

$$A_g = \int R\sqrt{-g}d^4x. \tag{10.1.19 b}$$

We thus find:

$$\delta A_g = \int [\delta R_{\mu\nu} g^{\mu\nu}\sqrt{-g} + R_{\mu\nu}\delta g^{\mu\nu} + R_{\mu\nu} g^{\mu\nu}\delta(\sqrt{-g})]\, d^4x. \tag{10.1.20}$$

By means of the identity:

$$\frac{\partial g}{\partial x^\alpha} = g g^{\mu\nu} g_{\mu\nu,\alpha}, \tag{10.1.21}$$

which arrises from the derivative of a determinant, we write:

$$\delta A_g = \int \left\{ \left[R_{\mu\nu} - \tfrac{1}{2} g_{\mu\nu} R\right] \delta g^{\mu\nu} + g^{\mu\nu}\delta R_{\mu\nu} \right\} \sqrt{-g}d^4x. \tag{10.1.22}$$

We leave for the end of this section the proof that the last term in the above integrand gives rise to a divergence integral which vanishes so that we are left with:

$$\delta A_g = \int \left\{ \left[R_{\mu\nu} - \tfrac{1}{2} g_{\mu\nu} R\right] \delta g^{\mu\nu} \right\} \sqrt{-g}d^4x = 0. \tag{10.1.23}$$

From (10.1.23) we find Einstein's field equations in the absence of the matter fields which are in turn given by (10.1.19 a); the combined result is given by:

$$G_{\mu\nu} = -\kappa T_{\mu\nu},$$

which coincides with the result of Chapter 3, Section 3.4. The constant κ arises here as the coupling constant between gravitation and matter fields: its precise value was obtained by comparison with Newtonian theory in the Chapter 3.

Proof of earlier claim:

$$\int \left\{ g^{\mu\nu}\delta R_{\mu\nu} \right\} \sqrt{-g}d^4x = 0. \tag{10.1.24}$$

We now show the above; from the definition of the Ricci tensor,

$$R_{ij} \equiv R^l_{ilj}. \tag{2.4.4}$$

So,

$$\delta R_{\mu\nu} = \Gamma^\alpha_{\mu\alpha}(\delta\Gamma^\beta_{\alpha\nu}) + \Gamma^\beta_{\alpha\nu}(\delta\Gamma^\alpha_{\mu\beta}) - \Gamma^\alpha_{\mu\nu}(\delta\Gamma^\beta_{\alpha\beta}) - \Gamma^\beta_{\alpha\beta}(\delta\Gamma^\alpha_{\mu\nu}) + (\delta\Gamma^\alpha_{\mu\alpha})_{;\nu} - (\delta\Gamma^\alpha_{\mu\alpha})_{;\alpha}.$$

It is easy to show that the differences between two given Γ 's are tensors, notwithstanding the fact that the individual Γ 's are not. So, the $\delta\Gamma$'s are tensors. From the covariant derivation rule and the above consideration, we write:

$$\delta R_{\mu\nu} = (\delta\Gamma^\alpha_{\mu\alpha})_{;\nu} - (\delta\Gamma^\alpha_{\mu\nu})_{;\alpha} \quad .$$

Equation (10.1.24) now becomes:

$$\int \left[(g^{\mu\nu}\delta\Gamma^{\alpha}_{\mu\alpha})_{;\nu} - (g^{\mu\nu}\delta\Gamma^{\alpha}_{\mu\nu})_{;\alpha} \right] \sqrt{-g} d^4x = \int \left[g^{\mu\nu}\delta\Gamma^{\alpha}_{\mu\alpha} - g^{\mu\alpha}\delta\Gamma^{\alpha}_{\mu\alpha} \right]_{;\nu} \sqrt{-g} d^4x =$$

$$= \int \{ \left[g^{\mu\nu}\delta\Gamma^{\alpha}_{\mu\alpha} - g^{\mu\alpha}\delta\Gamma^{\alpha}_{\mu\nu} \right] \sqrt{-g} \}_{,\nu} d^4x.$$

The above "volume" integral of an ordinary divergence transforms into a surface integral; as the variations $\delta\Gamma$'s, which are arbitrary and by hypothesis are null on the boundary, make the result of the integration equal to zero (Q.E.D).

10.2. Brans-Dicke Cosmology with Variable Lambda

To my recollection, except for Endo and Fukui's paper (1977), the physicist Orfeu Bertolami, in a different context, was the first one to treat the lambda problem by means of a variation of the "constant" with the inverse square of the age of the Universe (Bertolami, 1986 ; 1986 b), in a Brans-Dicke cosmological model with variable lambda, but in only two specific phases: $p = 0$ and $p = \frac{1}{3}\rho$ which represented respectively the dust case, and the radiation dominated Universe. Berman, (1990) studied the same problem under the general assumption of a perfect gas equation of state, for a perfect fluid, thus generalizing Bertolami's paper.

We shall show below that an accelerated Universe can be the outcome, for the Bertolami-Berman solution with a power law for the scale-factor, in a flat constant deceleration parameter model, with $0 > q > -1$. (Non flat solutions are also possible). Indeed, this model could also apply for a power law inflationary scenario, with $1 >> m \gtrapprox 0$. (Remember that $m = q + 1$). Nonetheless, exponential inflation can only occur if $\Lambda =$ constant, in this model.

We outline now the Bertolami-Berman model. The action is taken such that Λ–term is related to the scalar field $\phi(t)$:

$$S = \int d^4x \sqrt{-g} \left\{ -\tfrac{1}{16\pi}\phi \left[R - w\tfrac{\phi_{,\alpha}\phi^{,\alpha}}{\phi^2} \right] - \tfrac{\phi}{8\pi}\Lambda(\phi) + \mathcal{L}_m \right\}, \qquad (10.2.1)$$

where $\mathcal{L}_m$ is the Lagrangian of all other fields and particles.

On performing variations according to $g_{\mu\nu}$ and $\phi(t)$, we find, along with Bertolami (1986):

$$G_{\mu\nu} = \tfrac{8\pi}{\phi} T_{\mu\nu} + \tfrac{w}{\phi^2} \left[\phi_{;\mu}\phi_{;\nu} - \tfrac{1}{2} g_{\mu\nu}\phi_{;\alpha}\phi^{;\alpha} \right] + \tfrac{1}{\varphi} \left[\phi_{;\mu;\nu} - g_{\mu\nu}\Box^2\phi \right] - \Lambda g_{\mu\nu} \; ; \qquad (10.2.2)$$

$$\Box^2\phi + \tfrac{4\Lambda\phi}{3+2w} - \tfrac{4\phi^2}{3+2w}\tfrac{\partial\Lambda}{\partial\phi} = \tfrac{8\pi}{3+2w} T \; ; \qquad (10.2.3)$$

$$T = T^{\lambda}_{\lambda} \; ; \qquad (10.2.4)$$

$$T_{\mu\nu} \equiv \left(\tfrac{2}{\sqrt{-g}} \right) \tfrac{\delta L}{\delta g^{\mu\nu}} \; ; \qquad (10.2.5)$$

and

$$\Box^2 \equiv g^{\mu\nu}\nabla_\mu\nabla_\nu.$$

Local energy-momentum conservation is imposed altogether:

$$T^{\mu\nu}_{;\nu} = 0 \ ; \text{ and,} \tag{10.2.5.a}$$

$$G^{\mu\nu}_{;\nu} = 0. \tag{10.2.5.b}$$

The equivalence principle is saved, because we require that the scalar field does not couple in the equations of motion. Here,

$$T_{\mu\nu} = -pg_{\mu\nu} + [p+\rho]U_\mu U_\nu, \text{ where,}$$

$$U_\mu \equiv (1,0,0,0).$$

With the Robertson-Walker's metric, and the perfect gas equation of state, Bertolami finds:

$$\frac{\ddot{\phi}}{\phi} + 3H\frac{\dot{\phi}}{\phi} - \frac{4\Lambda}{3+2w} + \frac{4\phi}{3+2w}\frac{\partial\Lambda}{\partial\phi} = \frac{8\pi}{(3+2w)\phi}(\rho - 3p). \tag{10.2.6}$$

With the above, and from (10.2.2), the only non-zero components being the diagonals in $G_{\mu\nu}$, and with,

$$G(t)\phi = 1, \tag{10.2.7}$$

we find:

$$\frac{k}{R^2} + H^2 = \frac{8\pi\rho}{3\phi} - \frac{\dot{\phi}}{\phi}H + \frac{w}{6}\frac{\dot{\phi}^2}{\phi^2} - \frac{\Lambda}{3}. \tag{10.2.8}$$

From equation (10.2.5a) we obtain:

$$\dot{\rho} = -3H(\rho + p). \tag{10.2.8.a}$$

All equations, except for the conservation of energy relation (10.2.8 a) result from the following substitutions:

$$\rho \to \rho + \frac{\Lambda\phi}{8\pi}, \tag{10.2.9 a}$$

$$p \to p - \frac{\Lambda\phi}{8\pi} + \frac{1}{6\pi}\phi^2\frac{\partial\Lambda}{\partial\phi}. \tag{10.2.9 b}$$

We shall now show how Berman obtained a general solution of the type:

$$p = \alpha\rho, \ (\alpha = \text{constant}).$$

Our solution will be:

$$\Lambda = At^{-2}, \tag{10.2.10}$$

$$\phi = St^n, \tag{10.2.11}$$

$$\rho = FSt^{n-2}, \tag{10.2.12}$$

$$R = (mDt)^{1/m}, \tag{10.2.13}$$

where A, S, F, m, D are constants, and the deceleration parameter is constant,

$$q = m - 1 \ .$$

We plug back the above solution into the equations (10.2.6), (10.2.7), (10.2.8) and (10.2.9), obtaining:

$$B = n(n-1),$$

$$\frac{\ddot{\phi}}{\phi} = Bt^{-2},$$

$$n = -\frac{2A}{C},$$

$$\dot{\Lambda}\frac{\phi}{\dot{\phi}} = Ct^{-2},$$

$$m = \frac{-3(1+\alpha)}{n-2}.$$

The set of conditions to be obeyed by the constants of the solution is given by:

$$\frac{1}{m^2} = \frac{8\pi F}{3} - \frac{n}{m} + \frac{wn^2}{6} + \frac{A}{3},$$

$$-8\pi nF + (wn - \frac{3}{m})[(n-1)n - n^2] - \frac{3}{m^2}(2-m)n - 2A + \frac{3}{m}[8\pi F + \frac{w}{2}n^2 - \frac{3n}{m} - A] = 0 \ .$$

It is easy to check that the above solution can accommodate the case:

$$0 \geqslant q > -1,$$

pertaining to the present accelerating Universe, with a proper adjustment among the various constants of the above solution. We could easily obtain also the two particular solutions ($\alpha = 0$) and ($\alpha = \frac{1}{3}$) obtained in the paper by Bertolami (1986).

When Λ = constant, Uehara and Kim (1982) found also some particular solutions while Berman (1989) found exponential inflationary solutions.

An alternative treatment for Λ variable Brans-Dicke equations is that of Endo and Fukui (1977).

The variational principle is the same as above, but we define the μ constant by the equation:

$$\Box^2\phi = \frac{8\pi}{2\omega+3}\mu T,$$

where, if $\mu = 1$, we retrieve B.D. original theory. The field equations now read:

$$H^2 + \frac{k}{R^2} = \frac{1}{3}\left[\frac{8\pi}{\phi}\rho + \Lambda\right] - H\frac{\dot{\phi}}{\phi} + \frac{\omega}{6}\left(\frac{\dot{\phi}}{\phi}\right)^2, \tag{10.2.14}$$

$$\frac{d}{dt}\left(\dot{\phi}R^3\right) = \frac{8\pi}{2\omega+3}\mu\left(\rho - 3p\right)R^3, \tag{10.2.15}$$

and,

$$\Lambda = \frac{8\pi}{4\phi}\left(\mu - 1\right)\left(\rho - 3p\right). \tag{10.2.16}$$

By the methods outlined above, for Bertolami's equations, we can find a new solution for $p = \alpha\rho$ models. In (Berman, 1990 a), the reader will find a static solution for this theory, remembering that for "static" we mean $\dot{R} = 0$; the scalar-field $\phi(t)$, and $\Lambda(t)$ are time-varying. We also found a solution for an exponentially inflationary phase; however, we obtained a constant Λ, as a result of the manipulations (we remind that ϕ is time-varying in this model).(Berman, 1989).

A particular case, with $A = 0$, represents the solution for cosmological models in the original Brans-Dicke theory with $\Lambda = 0$. The seminal treatment is that of Weinberg (1972). Another author, A. Beesham (1986), also discussed the same Λ - form as in Bertolami's paper, but in other context. It is very easy to find also from the above, a solution with $k \neq 0$ (curved Universe), in the case $m = 1$ (see Berman and Som, 1989).

10.3. Model with Time Varying Fine Structure "Constant" - (1)

At the end of 20th century, (Webb et al 1999), and the beginning of this century, (Webb et al 2001), there were reports about the possible variation of the fine structure constant with the age of the Universe. Berman and Trevisan (2001) wrote three papers published in the site www.arXiv.org dealing with that subject. Berman has since found additional results which include a time-varying Λ - term, while in Berman and Trevisan's papers only in one case the Λ - term was present (see Section 10.4). Their findings are the subject of this Section and the two next ones.

Webb and collaborators provided experimental data on quasars that span 23% to 87% of the age of the Universe, finding deviation from the average, in the fine structure constant, given by $\frac{\triangle\alpha}{\alpha} \cong -0,72\text{x}10^{-5}$. In SI units, the fine structure constant α is given by:

$$\alpha \equiv \frac{e^2}{2\varepsilon_0 hc}, \tag{10.3.1}$$

where, e, ε_0,and h stands respectively for the charge of the electron, the electric permittivity and Planck's constant.

Due to the fact that α is defined by other constants, one can ask which is the constant that provokes the α variation. Another interesting remark is that this discovery is related to micro and macro phenomena, and can make the link between Quantum and Classical theories of gravity. Bekenstein (1982) proposed a theory with varying "e". An alternative theory is the varying speed of light (Moffat, 1993; Albrecht and Magueijo, 1999; Barrow, 1999; Berman and Trevisan, 2001). The latter will be discussed in the next two Sections, first in the context of the Generalized Large Number Hypothesis (GLNH) and then for a Jordan-Brans-Dicke framework.

We now present the scenario of time varying ε_0 while "c", "h" and "e" are strictly constant. We resort to GLNH. One could claim that a specific gravitational theory is needed in order to make the analysis correct; however, it is not certain, in present time, which is the correct theory of gravity that is to be adopted so our naive analysis can help in visualizing what is going on. If overdots stand for time derivatives, from (10.3.1) we find that:

$$\frac{\dot{\alpha}}{\alpha} = -\frac{\dot{\varepsilon}_0}{\varepsilon_0}. \tag{10.3.2}$$

We suppose now, tentatively, that ε_0 varies with a power law of time, say:

$$\varepsilon_0 = At^n, \tag{10.3.3}$$

with $A, n =$ constants. Then:

$$\frac{\dot{\alpha}}{\alpha} = -nt^{-1}. \tag{10.3.4}$$

With a constant deceleration parameter model we again find:

$$q = -\frac{\ddot{R}R}{\dot{R}^2} = m - 1 \ ; \tag{10.3.5}$$

$$H = \frac{\dot{R}}{R} = \frac{1}{mt} \tag{10.3.6}$$

$$R = R(t) = (mDt)^{1/m}, \tag{10.3.6.a}$$

where m, $D =$ constants.

We remark that the theory of constant deceleration parameters defined by the above three formulae independs on the particular gravitational theory that is considered. On the other hand, the experimental value found by Webb et al, may be interpreted as:

$$\frac{\Delta\alpha}{\alpha\Delta t} \simeq -\frac{0.72 \text{ x } 10^{-5}}{(0.87-0.23)t} \simeq -1.1 \text{ x } 10^{-5} Hm. \tag{10.3.7}$$

For relations (10.3.4) and (10.3.7) we find:

$$n \simeq 1.1 \text{ x } 10^{-5}. \tag{10.3.8}$$

This is the way in which the permittivity has to vary in our framework. From electromagnetism, we know that:

$$c = (\varepsilon_0 \mu_0)^{-1/2}. \tag{10.3.9}$$

In order to keep $c =$ constant, we must vary μ_0 inversely as ε_0. We resort now to GLNH:

$$\frac{cH^{-1}}{4\pi\varepsilon_0 (\frac{e^2}{m_e c^2})} \cong \sqrt{N}. \tag{10.3.10}$$

$$\frac{e^2}{4\pi\varepsilon_0 G m_p m_e} \cong \sqrt{N}. \tag{10.3.11}$$

$$\frac{\rho (cH^{-1})^3}{m_p} \cong N. \tag{10.3.12}$$

$$ch(m_p m_e/\Lambda)^{1/2} \cong \sqrt{N} \quad ; \qquad (10.3.13)$$

where m_e, m_p,and N stands respectively for electron's and proton's mass and the total number of nucleons in the Universe, respectively. We now consider that N increases with the age of the Universe, and find the time variations of N, G, ρ and Λ that are coherent with the hypotheses (10.3.3); (10.3.6) and (10.3.8):

$$N \propto t^{1.99998}. \qquad (10.3.14)$$

$$G \propto t^{-1.0}. \qquad (10.3.15)$$

$$\rho \propto t^{-1.00002}. \qquad (10.3.16)$$

$$\Lambda \propto t^{-1.99998}. \qquad (10.3.17)$$

We hope that the next generation of experimentalists will provide checks on the above results. The time variation for Λ is very close to the law advocated by us in earlier Chapters. It is doubtful whether, in the near future, the powers in (10.3.16) and (10.3.17), could be experimentally checked against the obvious laws:

$$\rho \propto t^{-1},$$
$$\Lambda \propto t^{-2}.$$

Nevertheless, it may happen that the variation law for α, and ε_0 could be of major importance in astrophysical or nuclear physics.

Prior work with a time varying ε_0 should be credited to Gomide (1976) who nevertheless worked with α = constant, in face of Bahcall and Schmidt's paper (1967). We also point out that the origin of $c = c(t)$ theories can be traced to Gomide's paper (1976) and that the $\dot{\alpha} \neq 0$ theories with variable speed of light were also considered several times later by Barrow and Magueijo (see for instance (1999)).

Confronting with observations for the time variation of the gravitation constant G, we define:

$$\frac{\dot{G}}{G} = \sigma H. \qquad (10.3.18)$$

Our result is, from (10.3.15):

$$\frac{\dot{G}}{G} \cong -1.0\, t^{-1} \cong -1.0\, H\ (1+q), \qquad (10.3.19)$$

where we have again used relations (10.3.6). This means that if we could have accurate measures of σ, we could estimate the deceleration parameter q. We refer to Will (1987; 1995) in order to mention that there is no specific and conclusive experimental value for σ. Lunar laser ranging and Viking radar measurements by Williams et al (1976) and Reasenberg (1983), put

$$|\sigma| < 0.6.$$

In Will (1995) it is stated that this two kinds of measurements give the best limits on $\dot{G}/G$. (See also Dickey (1994)). This means that:

$$-0.4 > q > -1.6, \tag{10.3.20}$$

for the present Universe, that would be accelerating, as supernovae results confirmed. (Perlmutter et al,1997 ; 1998; Garnavich et al, 1998; Schmidt, 1998; Riess, 1998; etc.).

We have shown that GLNH, fine structure constant time variation, the accelerating Universe, the variable lambda or variable permittivity, are all coherent among them. The present constant deceleration parameter that points out towards an accelerating Universe must not be taken to deny that, other values of q were possible ten billion years ago as it seems from an old Supernova explosion evidence (reported by Schwarzschild, 2001).

One final comment remains necessary: it will be the task of a Superunification theory to explain such or other time variation of fundamental constants. In this Section and in the next, where we employ the GLNH hypothesis tentatively, we just hint towards future better analyses.

10.4. Model with Time Varying Fine Structure "Constant" - (2)

We refer to the preceding Section 10.3 for information on the experimental time variability of α. One can ask whether this variation could be caused by a time-varying speed of light: $c = c(t)$. Gomide (1976) has studied such a case, which was later revived by Barrow (1998 ; 1999 ; 1997) and Barrow and Magueijo (1999). This could explain also Supernovae observations. We refer to their papers for further information. Our framework now will be an estimate made through Berman-Dirac GLNH. We write again the main formulae for constant deceleration parameters, as derived originally in the literature by Berman (1983) and Berman and Gomide (1988):

$$H = \frac{\dot{R}}{R} = \frac{1}{mt} = \frac{1}{1+q} t^{-1}. \tag{10.4.1}$$

$$q = -\frac{\ddot{R}R}{\dot{R}^2} = m - 1. \tag{10.4.2}$$

$$R = (mDt)^{1/m}. \quad (m \neq 0), \tag{10.4.3}$$

where R is the scale-factor in Robertson-Walker's metric:

$$ds^2 = dt^2 - \frac{R^2(t)}{(1+\frac{kr^2}{4})^2}(dx^2 + dy^2 + dz^2). \tag{10.4.4}$$

We express now Webb et al's experimental result as:

$$\left(\frac{\dot{\alpha}}{\alpha}\right)_{\exp} \simeq -1.1\,x\,10^{-5}\,H\,(1+q). \tag{10.4.5}$$

We define the fine structure "constant" as in Barrow's papers cited above:

$$\alpha \equiv \frac{e^2}{\hbar\, c(t)}. \tag{10.4.6}$$

From (10.4.6) we find:

$$\frac{\dot{\alpha}}{\alpha} = -\frac{\dot{c}}{c}. \qquad (10.4.7)$$

Again, we suppose that the speed of light varies with a power law of time:

$$c = At^n \qquad (A = \text{constant}). \qquad (10.4.8)$$

From the above experimental value we find:

$$n = 1.1 \text{ x } 10^{-5}. \qquad (10.4.9)$$

We see that the speed of light varies slowly with the age of the Universe, as we shall also find in next Section. Now let us consider Berman-Dirac's Generalized Large Number Hypothesis (GLNH): we call N the number of nucleons in the Universe; our four relations remain:

$$\frac{cH^{-1}}{\left(\frac{e^2}{m_e c^2}\right)} \cong \sqrt{N}. \qquad (10.4.10)$$

$$\frac{e^2}{Gm_p m_e} \cong \sqrt{N}. \qquad (10.4.11)$$

$$\frac{\rho(cH^{-1})^3}{m_p} \cong N. \qquad (10.4.12)$$

$$ch(m_p m_e/\Lambda)^{1/2} \cong \sqrt{N}. \qquad (10.4.13)$$

By means of (10.4.1) and the above four relations, we obtain the following set of solutions:

$$\rho \propto t^{-2}\sqrt{N}. \qquad (10.4.13\text{ a})$$

$$G \propto N^{-1/2}. \qquad (10.4.14)$$

$$c \propto N^{\frac{1}{6}} t^{-\frac{1}{3}}. \qquad (10.4.15)$$

In order to accommodate (10.4.15) with (10.4.8) and (10.4.9), we impose:

$$N \propto t^{2.0001}, \qquad (10.4.16)$$

and then:

$$G \propto t^{-1.00005}. \qquad (10.4.17\text{ a})$$

$$\Lambda \propto t^{-2.0001}. \qquad (10.4.17\text{ b})$$

$$\rho \propto t^{-0.99995}. \qquad (10.4.18)$$

This is our solution, based on Berman-Dirac's GLNH. A similar incomplete solution was already published by Berman and Trevisan (2001). The same doubts expressed in last Section, on the viability of experimental measurements in order to check the power exponents in the above laws, are present here.

A close analysis shows how the deceleration parameter range is situated. Both from Lunar laser ranging and Viking radar measurements by Williams et al (1976) and Reasenberg (1983) we find:

$$\frac{\dot{G}}{G} = \sigma H \qquad \text{with} \quad |\sigma| < 0.6.$$

Will (1987; 1995) and Dickey (1994) comment that these two kinds of measurements give the best limit on $|\sigma|$. Because in our model we have:

$$\frac{\dot{G}}{G} \cong 1.0\, t^{-1} \cong 1.0\, H\, (1+q), \tag{10.4.19}$$

we find:

$$-0.4 > q > -1.6. \tag{10.4.20}$$

The above result is satisfactory, for a negative "q" agrees with Supernovae observations.

We have thus shown that $\frac{\Delta\alpha}{\alpha}$ should really be negative, for a positive value, could mean a positive deceleration parameter. As a bonus we found possible laws of variation for N, G, ρ, and Λ. The Λ–term time variation is also very close and even, practically indistinguishable, from our last Chapters law of variation $\Lambda \propto t^{-2}$.

It is clear that in this Section's model, the electric permittivity of the vacuum, along with its magnetic permeability, and also Planck's constant are really constant here. We point out again, that in the long run, it will be only when a Superunification theory becomes available, that our different models could be discarded, (hopefully) but one.

10.5. Model with Time Varying Fine Structure "Constant" - (3)

Instead of tackling the latest developments of Physics in the present century which are the acceleration of the present Universe and the fine structure time variation, by means of GLNH, we shall try now a different approach, based on the work of the well known physicists John D. Barrow (1996 and 1999) and Barrow and Magueijo (1999) and Barrow and Parsons (1997): it is based on a Jordan-Brans-Dicke model along with variable speed of light. We will generalize their approach, by including a time varying cosmological term [Berman and Trevisan, 2001 c studied the same model without cosmological term].

We again go ahead as in Berman (1983) and Berman and Gomide (1988) with constant q models:

$$q = -\frac{\ddot{R}R}{\dot{R}^2} = m - 1 = \text{constant}. \tag{10.5.1}$$

We define the fine structure constant as usual:

$$\alpha \equiv \frac{e^2}{\frac{h}{2\pi}c}. \tag{10.5.2}$$

We recall the experimental time variation gathered by Webb et al (1999), (2001):

$$\frac{\Delta\alpha}{\alpha} \simeq -0.72 \text{ x } 10^{-5}. \tag{10.5.3}$$

$$\Delta t = (0.87 - 0.23)t. \tag{10.5.4}$$

(The above refers for observations relating to 23% to 87% of the age of the Universe).

The α variation will be attributed solely to a time varying speed of light, $c = c(t)$; it does not seem that the charge of the electron or Planck's constant have a large scale role in the Universe.

Webb et al pointed out that a common property of unified theories, when applied to Cosmology is that they predict a space and time dependence of the coupling constants. Let us suppose then, that G is also time varying; the natural framework we need is JBD (Jordan-Brans-Dicke) theory with varying speed of light as Barrow, Magueijo and Parsons have been studying. Our equations read, when one adds the Λ - term for a flat Universe:

$$H^2 = \frac{8\pi\rho}{\phi} - \frac{\dot{\phi}}{\phi}H + \frac{w}{6}\frac{\dot{\phi}^2}{\phi^2} - \frac{kc^2(t)}{R^2} - \frac{\Lambda}{3}. \tag{10.5.5}$$

$$\frac{\ddot{\phi}}{\phi} + \frac{3H\dot{\phi}}{\phi} = \frac{-8\pi\phi^{-1}}{3+2w}\left(\rho - \frac{3p}{c^2(t)}\right) - \frac{4\Lambda}{3+2w} + \frac{4\phi}{3+2w}\frac{\partial\Lambda}{\partial\phi}. \tag{10.5.7}$$

Unlike Barrow, we shall not impose a definite equation of state; the values for the energy density ρ and the cosmic pressure p will arise naturally in our solution.

From the theory of Constant Deceleration Parameter we have:

$$H = (mt)^{-1}. \tag{10.5.8}$$

Introducing the arbitrary constants A, B, γ, F, C and S, we find the following solution for equations (10.5.5), (10.5.6) and (10.5.7):

$$R = (mDt)^{\frac{1}{m}}. \tag{10.5.9}$$

$$\phi = At^n. \tag{10.5.10}$$

$$\rho = ABt^{n-2}. \tag{10.5.11}$$

$$p = Ft^{\gamma-4+2/m}. \tag{10.5.12}$$

$$c = Ct^{(1/m)-1}. \tag{10.5.13}$$

$$\Lambda = St^{-2}. \tag{10.5.14}$$

Let us now find our theoretical result, from (10.5.13) and the definition of α:

$$\frac{\dot{\alpha}}{\alpha} = Hq. \tag{10.5.15}$$

We compare now with the experimental result, namely:

$$\frac{\Delta\alpha}{\alpha\Delta t} \cong \frac{\Delta\alpha}{\alpha\left[0.64\frac{H^{-1}}{m}\right]} \cong -\frac{0.72}{0.64}10^{-5}Hm \cong -1.1\,x\,10^{-5}Hm. \qquad (10.5.16)$$

From the comparison, we find an estimate for the deceleration parameter:

$$m \cong 0.99999$$

or

$$q \cong -1.1 \text{ x } 10^{-5}. \qquad (10.5.17)$$

We see that the deceleration parameter is slightly negative, but very close to zero. The speed of light, likewise in the last Section, increases slowly with time. The cosmological constant varies like usual.

Suggested reading:

The papers by John D. Barrow cited in the References at the end of this book.

10.6. The Machian Universe Revisited

The international media, has published news about a discovery, by A. Riess and collaborators, that the Universe has been Λ – driven, for the most part of its history. We have, in previous Chapters, presented the Machian Universe, in such a way, that the above fact, was evident.

Barbour and Pfister(1995), listed several possible meanings generally attributed to a Machian Universe, among others, as: a) the problem of motion definition; b) the determination of inertial frames; c) the interactions between masses; d) the generation of inertia; e) the induction of inertial forces by means of accelerations, in analogy with electromagnetism; f) the requirement that the metric tensor should be determined by matter; or by matter and geometric degrees of freedom for gravity; g) the cosmic derivation of inertial mass by means of the ADM Hamiltonian; h) the generalization of Special Relativity when gravitation is present; i) the requirement of general covariance of physical laws; j) the existence, or not, of boundary conditions for the Universe; or the existence or inexistence of matter-free singularity-free solutions in Cosmology; k) the requirement of physical causality in Nature; l) the prediction of dragging effects due to the distribution of masses; m) the antidote to Gödel-type Universes; n) the explanation of the origin of Brans-Dicke relation; o) the selection among possible cosmological models; p) the non-existence of an absolute dynamical theory; etc.

We now shall propose that Mach's Principle, means a zero-total energy Universe(Feynman, 1962-3). Berman(2006; 2006a), has shown this meaning of Mach's Principle without considering a rotating Universe. We now extend the model, in order to include the spin of the Universe, and we replace Brans-Dicke traditional relation, $\frac{GM}{c^2R} \sim 1$, with four different relations, which we shall call the Brans-Dicke relations for mass, lambda, spin, and temperature of the Universe

We shall consider a "large" sphere, with mass M, radius R, spin L, and endowed with a cosmological term Λ, which causes the existence of an energy density $\frac{\Lambda}{\kappa}$, where $\kappa = \frac{8\pi G}{c^2}$. We now calculate the total energy E of this distribution:

$$E = E_i + E_g + E_L + E_\Lambda + E_R, \tag{10.6.1}$$

where $E_i = Mc^2$, stands for the inertial (Special Relativistic) energy; $E_g \cong -\frac{GM^2}{R}$ (the Newtonian gravitational potential self-energy); $E_L \cong \frac{L^2}{MR^2}$ the Newtonian rotational energy; $E_\Lambda \cong \frac{\Lambda R^3}{6G}$ (the cosmological "constant" energy contained within the sphere), and $E_R = aT^4$, where a is a constant, and T stands for absolute temperature, while E_R represents radiational energy.

If we impose that the total energy is equal to zero, i.e., $E = 0$, we obtain from (10.6.1):

$$\frac{GM}{c^2R} - \frac{L^2}{M^2c^2R^2} - \frac{\Lambda R^3}{6GMc^2} - \frac{E_R}{Mc^2} \cong 1. \tag{10.6.2}$$

As relation(10.6.2) above should be valid for the whole Universe, and not only for a specific instant of time, in the life of the Universe, and if this is not a coincidental relation, we can solve this equation by imposing that:

$$\frac{GM}{c^2R} = \gamma_G \sim 1, \text{ (10.6.3) } \frac{L}{McR} = \gamma_L \sim 1; \text{(10.6.4) } \frac{E_R}{Mc^2} = \frac{\rho_R R^3}{Mc^2} = \frac{aT^4R^3}{Mc^2} = \gamma_R \sim 1, \text{ (10.6.4a)}$$

and,

$$\frac{\Lambda R^3}{6GMc^2} = \gamma_\Lambda \sim 1, \tag{10.6.5}$$

subject to the condition,

$$\gamma_G - \gamma_L^2 - \gamma_\Lambda - \gamma_R \cong 1, \tag{10.6.6}$$

where the $\gamma' s$ are constants having order "one".

We now propose the following generalized Brans-Dicke relations, for gravitation, spin, radiation and cosmological "constant":

$$\frac{GM}{c^2R} = \gamma_G \sim 1, \tag{10.6.3}$$

$$\frac{GL}{c^3R^2} = \gamma_G \cdot \gamma_L \sim 1, \tag{10.6.7}$$

$$\frac{aGT^4R^2}{c^4} = \gamma_R \cdot \gamma_G \sim 1, \tag{10.6.7a}$$

and,

$$\frac{\Lambda R^2}{6c^4} = \gamma_\Lambda \cdot \gamma_G \sim 1. \tag{10.6.8}$$

The reader should note that we have termed Λ as a "constant", but it is clear from the above, that in an expanding Universe, $\Lambda \propto R^{-2}$, so that Λ is a variable term. We also notice that $R \propto M$, and $L \propto R^2$, and $R \propto T^{-2}$, so that, we should have, $\rho_R \propto R^{-2}$. The T^{-2}

dependence on R, was dealt, earlier, for non-relativistic decoupled massive species (Kolb and Turner, 1990).

The B.D. relation for spin, has been derived, earlier, on a heuristic procedure, which consists on the simple hypothesis that L should obey a similar relation as M (Sabbata and Sivaram, 1994). The first authors to propose the above R^{-2} dependence for Λ were Chen and Wu (1990), under the hypothesis that Λ should not depend on Planck's constant, because the cosmological "constant" is the Classical Physics response to otherwise Quantum effects that originated the initial energy of the vacuum. Berman, as well as Berman and Som, have examined, along with other authors, a time dependence for Λ(see for example, Berman, 1991; 1991b).

It must be remarked, that our proposed law (10.6.3), is a radical departure from the original Brans-Dicke (Brans and Dicke, 1961) relation, which was an approximate one, while our present hypothesis implies that $R \propto M$. With the present hypothesis, one can show, that independently of the particular gravitational theory taken as valid, the energy density of the Universe obeys a R^{-2} dependence (see Berman, 2006; 2006a; Berman and Marinho, 2001). In fact, from the definition of matter (inertial) energy density:

$$\rho = \frac{M}{V}, \tag{10.6.9}$$

while,

$$V = \alpha R^3, \quad (\alpha = \text{constant}) \tag{10.6.10}$$

where ρ and V stand for energy density and tridimensional volume, we find:

$$\rho = \left[\frac{\gamma_G}{G\alpha}\right] R^{-2}. \tag{10.6.11}$$

If we apply the above relation, for Planck's and the present Universe, we find:

$$\frac{\rho}{\rho_{Pl}} = \left[\frac{R}{R_{Pl}}\right]^{-2}. \tag{10.6.12}$$

If we substitute the known values for Planck's quantities, while we take for the present Universe, $R \cong 10^{28}$ cm, we find a reasonable result for the present energy density. This shows that our result (relation 10.6.11), has to be given credit. For the energy density of the Λ - term, we shall have $\rho_\Lambda = \frac{\Lambda}{\kappa} \propto R^{-2}$, and also for the rotational energy, we have $\rho_L = \frac{L^2}{c^2 M^2 R^4} \propto R^{-2}$; we have also shown above that the radiation energy density is also proportional to R^{-2}. We conclude that all forms of energy densities are R^{-2} - dependent, provided that the generalized Brans-Dicke relations, for gravitation, spin, cosmological "constant" and radiation, obey the conditions (10.6.3), (10.6.7), (10.6.7a) and (10.6.8) above.

The entropy of the Machian Universe, as described above, is not constant, but increases with time:

$$S = sR^3,$$

where, S and s represent total entropy, and entropy density, respectively for the Universe. According to the above formulae, we have:

$$S \propto sR^3 = \frac{\rho_R}{T} R^3 \propto T^3 R^3 \propto R^{\frac{3}{2}} \propto M^{\frac{3}{2}}, \qquad (10.6.12a)$$

taken care of the relations given earlier, i.e., $R \propto T^{-2}$, and (10.6.3) above.

We have thus shown that S increases with time, likewise the scale factor to the power $\frac{3}{2}$.This Machian property stands opposite to the usual assumptions in theoretical studies that, by considering $S =$ constant, found $R \propto T^{-1}$, which as we have shown, is incorrect in the Machian picture.

This solves the "loss of information paradox", for black holes, because this body fills a bounded region of space, limited by the horizon. As the Universe is isotropic and homogeneous, each bounded region of it, must share the growth in entropy of the whole Universe, because the mass of black hole, being part of the Universe, increases likewise, with the mass of the Universe. Remember: a Classical b.h. does not evaporate.

It should be remembered that the origin of Planck's quantities, say, for length, time, density and mass, were obtained by means of dimensional combinations among the constants for macrophysics (G for gravitation and c for electromagnetism) and for Quantum Physics (Planck's constant $\frac{h}{2\pi}$). Analogously, if we would demand a dimensionally correct Planck's spin, obviously we would find,

$$L_{Pl} = \frac{h}{2\pi}. \qquad (10.6.13)$$

From Brans-Dicke relation for spin, we now can obtain the present angular momentum of the Universe,

$$L = L_{Pl}\left[\frac{R}{R_{Pl}}\right]^2 \cong 10^{120}\left(\frac{h}{2\pi}\right) = 10^{93} \quad g\ cm^2\ s^{-1}. \qquad (10.6.14)$$

This estimate was also made by Sabbata and Sivaram(1994), based on heuristic considerations(see also Sabbata and Gasperini, 1979).

If we employ, for the cosmological "constant" Planck's value, Λ_{Pl},

$$\Lambda_{Pl} \cong R_{Pl}^{-2}, \qquad (10.6.15)$$

then, we shall find, in close agreement with the present value estimate for Λ (as found by recent supernovae observations), by means of the third Brans-Dicke relation:

$$\Lambda = \Lambda_{Pl}\left[\frac{R_{Pl}}{R}\right]^{-2}. \qquad (10.6.16)$$

Sabbata and Gasperini(1979), have calculated the angular speed of the Universe, for the present Universe. Though they mixed their calculations with some results obtained from Dirac's LNH (Large Number Hypothesis), including a time variation for the gravitational "constant", we now show that, if we take for granted that $G =$ constant, and by means of the generalized Brans-Dicke relations we find, by considering a rigid rotating Universe, whereby:

$$L = MR^2\omega, \qquad (10.6.17)$$

so that,

$M\omega =$ constant, (because $L \propto R^2$ as we have shown earlier), we shall have:

$$\omega_{Pl} = \frac{c}{R_{Pl}} = 2 \text{ x } 10^{43} \ s^{-1}, \qquad (10.6.18)$$

and, for the present,

$$\omega = \frac{c}{R} \cong 3 \text{ x } 10^{-18} \ s^{-1}. \qquad (10.6.19)$$

It must be pointed out, that the result (10.6.18) seems to us that was not published elsewhere, up to now. This is, however, the first time that the above results are obtained by means of the zero-total energy hypothesis for the Universe.

We remember that the Pioneers' spaceships were found to have an anomalous acceleration, of unexplained origin; Berman(2006 e) has shown that the centripetal acceleration $a = -\omega^2 R$, in which we take the numerical values from (10.6.19), result in the sought value; we may say that, in the Machian Universe, between any two points for, an "observer" and an "observed", there is a relative acceleration, with the above value, which is ubiquitous.

This is why we attribute this hypothesis to a Machian Universe; indeed, we believe that we can identify Mach's Principle, with this hypothesis. Then, the entropy of the Universe grows with increasing time, because it is expanding, and so does the entropy of any black hole.

If the black hole, even in a Classical Thermodynamical picture, is seen as a black body, its energy density would be given, in terms of the absolute temperature $T_b h$, by,

$$\rho_r = \sigma T^4, \quad (\sigma = \text{constant}), \qquad (10.6.20)$$

while on the other hand, the energy densities vary with R^2 , we find, that, from the thermodynamical black body entropy law (Sears and Salinger, 1975):

$$S_{bh} \propto R_s^3 T_{bh}^3, \qquad (10.6.21)$$

and then,

$$R_S T_{bh}^2 = constant, \qquad (10.6.22)$$

while,

$$S_{bh} \propto R_S^{3/2}. \qquad (10.6.23)$$

We see that the Universe and black holes, obey similar entropy laws. However, in a Classical viewpoint, the temperature of a black hole, is just an attached label, and does not imply in evaporation. That is why for a fixed horizon, the black hole has fixed entropy, in a local sense, when we may forget locally that the Universe expands.

Casimir Effect

I want to point out, that the Machian Principle, must not be restricted to the gravitational interaction. It is known that when two close and parallel metallic plates are set in the vacuum, they experience an attractive force towards each other. This is called the Casimir effect, and is due to the existence of virtual pairs of particles's production in the vacuum; of course, outside the region between the plates, there are more virtual pairs being created, then inside. Therefore, the exterior pressure on the plates, is larger than the interior one, hence the verified attraction.

From the above description, we find that the global distribution of virtual pairs's formation, gives rise to local effects. This is a Machian non-gravitational effect, in my opinion.

10.7. Dark Matter as Neutrinos with Time-Varying Mass

This Section is based on research done in collaboration with F.M. Gomide. The subject of mass-varying neutrinos, has been very recently given attention (Horvat, 2005; Fardon et al., 2003; Kaplan, 2004). But, first, let us make an historical review of the neutrino research.

In 1930 the Austrian theoretical physicist Wolfgang Pauli, Nobel Prize due to his research on the electron spin, verified in the processes of electron emission from neutrons, that the energy-momentum balance was violated. In other terms: the much solid principle of energy-momentum conservation of special relativity seemed to loose its preeminence at the basis of every physical theory. Well something highly perplexing, for in spite of being universally applied to all known high energy processes, it could not be corroborated in this special process of electron emission from neutrons, the so called Beta emission process. A painful problem of course, but brilliantly solved through Pauli's revolutionary hypothesis and Enrico Fermi's theory, in 1934, (Schweber, 2002). Pauli postulated the existence of a new particle, the neutrino(Franklin, 2000) that was responsible for the disappearance of energy-momentum in the neutron decay (beta emission). Fermi forwarded the neutrino theory wherein it is deduced that this new postulated neutral particle has a probability of interaction with other particles, 10 to power 22 times smaller than what was known then. Therefore the neutrino acted as if a phantom particle that could not be easily detected. Only in 1953, twenty years after Fermi's theory, was the particle to emerge to existence, thanks to the experiments done by two American experimental physicists, Frederick Reines and Clyde Cowan. They used then a new detection technique that afforded the possibility for detection of an effect predicted by Fermi. It was the technique based on a different type of particle detector, the scintillation counter. Being faithful to a fundamental physical principle, the energy-momentum conservation, Pauli and Fermi were responsible for one of the most important scientific discovery of the XXth century. Fermi's predictions from his theory were widely verified by the community of high-energy physicists, such that nobody doubted the existence of the putative neutrino. Frederick Reines who received the Nobel Prize in 1995 made the following important epistemological comment:

> "It must be recognized, however, that independent of the observation of a 'free neutrino' interaction with matter, the Fermi theory was so attractive in its explanation of beta decay that belief in the neutrino as a 'real' entity was general".(Franklin, 2000)

Reines' comment is a remarkable epistemological judgement in so far as to view certain experimental evidence of a theory, not something clinching or crucial. Fermi's theory offered a galaxy of effects in high-energy physics of so important level, that the putative neutrino could be kept as such for 20 years without bothering physicists. Reiner and Cowan's discovery of the free neutrino was not a crucial test of a theory, but an experimental success after a history of theoretical and other experimental evidences. We are going to deal afterwards on the question of clinching experimental proofs. An analogous situation we are going to point out in the history of the heliocentric theory, accepted by every astronomer of the XVIIth century without the "clinching proof" of stellar parallax obtained by Bessel on the XIXth century. Let us remind that Enrico Fermi's theory of 1934 is considered the birth of quantum field theory(Schweber, 2002), which is cornerstone in elementary particles theory. Fermi received the Nobel in 1938.

The present theory of material reality's offers three families of elementary particles:

a) 6 quarks and 6 antiquarks.

b) Leptons and antileptons: electron, muon, tau-particle, neutrino-e, neutrino-mu and neutrino-tau.

c) Gauge bosons: photon, W+, W-, Z, and Gluon. Being bosons they have no antiparticles' counterpart.

Let us add some considerations as to the three neutrinos. The neutrino developed in Fermi's theory and detected by Reines and Cowan was called by Feinberg in 1958, neutrino-e, since it was related to the beta decay of the neutron. Besides Feinberg predicted that the neutrino generated by the muon decay should be different and it was then named neutrino-mu. Through an experiment done in 1962 by Leon Lederman, Schwartz, and Steinberger, the prediction was brilliantly corroborated(Lederman, 1989). In 21/7/2000 a group of 54 American, Japanese, Korean, and Greek physicists at Fermilab, announced the detection of the neutrino-tau.(Fermilab, 2000) The neutrino-tau results from the decay of the lepton tau particle. Thus, from 1953 to 2000 the 3 neutrinos predicted in the theoretical structure had been detected.

From 1967 to 1998 several experiments were done in order to measure the neutrino flux produced in the Sun. As we know the thermonuclear reactions inside our star, would produce abundant torrents of neutrinos-e. The first measurements done in 1967 by Ray Davies offered a perplexing result: the flux measured had been about 1/3 of the predicted value(Science, 1992; 1992 a). In 1969 the Italian physicist Pontecorvo suggested a hypothesis that would account for the missing Solar neutrinos: en route to Earth part of them would oscillate to a different type of neutrinos. This hypothesis was endorsed in 1985 by two Russian physicists Miteyev and Smirnov: from the Sun to Earth a substantial part of neutrinos-e would transform into neutrinos-mu or tau(Schwarzschild, 1992).

According to the standard model of elementary particles theory, the property of oscillation of a neutrino into another type presupposes the existence of mass. Hence the old assumption of the massless neutrino should be disposed off. A team of 300 American and Japanese physicists at Takayma proceeded to an experiment with the huge detector Super-Kamiokande in order to detect oscillation of neutrino-mu into neutrino-tau in the flux of secondary cosmic rays. In spite of the cautious language of Yoshiro Suzuki of the University of Tokyo, the results strongly favour the oscillation of neutrinos-mu(Schwarzschild, 1998). These findings with the Super-Kamiokande are regarded as compelling but not definitive(Science, 1999). More recently, Tsuyoshi Nakaya of Kyoto University announced in June, 2004 at a meeting in Paris, a reassuring confirmation of the oscillation mechanism, that is: strong evidence of oscillation of neutrino-mu into neutrino tau(Physics Today, 2004). The square masses of the neutrino eigenstates were measured from the Super-Kamiokande data and were in agreement with the oscillation parameters from atmospheric data.

Notwithstanding the successes obtained, new theoretical problems are being tackled in the domain of the physics of neutrinos(Masiero et al., 2004). Well, there is no end in the quest for knowledge and understanding of reality: problems solved, open the doors for new problems.

MACHIAN RELATIONS FOR DARK MATTER ENERGY

It has been asserted, that 67% of the energy density of the Universe, is due to a cosmological "constant" energy. The remnant energy density is fractionated in two parts: 5% as visible mass and 28% as dark matter. Let us suppose that dark matter is constituted by neutrinos with non-zero rest mass. (Berman, 2006; 2006 b) has suggested, (see also Section 10.6) that, if Mach's principle is understood as meaning that the total energy of the Universe is null, and if each particular energy contribution to the total energy density, has constant participation during the whole history of the Universe, one may obtain different Machian relations. These Machian relations, of which, the Brans-Dicke (1961) relation is a particular case, should not, according to Berman, be viewed as just coincidental with the present Universe. Suppose that the total energy is given by:

$$E = Mc^2 - \frac{GM^2}{2R} + 4\pi\Lambda\frac{R^3}{3\kappa} + \frac{L^2}{MR^2}, \tag{10.7.1}$$

where the four terms to the right of relation (10.7.1) represent respectively the inertial, gravitational, cosmological constant's and rotational energies.

When we impose,

$$E = 0, \tag{10.7.2}$$

we obtain:

$$\frac{GM}{2c^2R} - \frac{4\pi}{3\kappa}\left[\frac{\Lambda R^3}{c^2M}\right] - \frac{L^2}{c^2M^2R^2} = 1. \tag{10.7.3}$$

If no one of the above terms will increase or decrease different than the others, we can solve equation (10.7.3) by imposing that:

$$\frac{GM}{2c^2R} = \gamma_1, \tag{10.7.4}$$

$$\frac{4\pi}{3\kappa}\left[\frac{\Lambda R^3}{c^2M}\right] = \gamma_2, \tag{10.7.5}$$

$$\frac{L^2}{c^2M^2R^2} = \gamma_3, \tag{10.7.6}$$

where the γ's obey the conditions:

1) $\gamma_i =$ constant $(i = 1,2,3)$, (10.7.7)

$$2)\quad \gamma_1 - \gamma_2 - \gamma_3 = 1. \tag{10.7.8}$$

It can be checked that, due to the original Brans-Dicke relation (Brans and Dicke, 1961),

$$\frac{GM}{c^2R} \sim 1, \tag{10.7.9}$$

and also because of the above arguments, all the γ's are of order "one". We thus obtain, according to Section 10.6, the following variation laws:

$$M \propto R\,; \tag{10.7.10}$$

$$L \propto R^2\,; \tag{10.7.11}$$

$$\Lambda \propto R^{-2}. \tag{10.7.12}$$

The Machian relations (10.7.4)(10.7.5)(10.7.6) have been noticed long ago, as approximate relations for the present Universe; the radical departure made by Berman, is contained in the fact that the γ's are constant during the lifespan of the Universe, and not only for the present time, so that relations (10.7.10)(10.7.11)(10.7.12), are valid during all times.

We now can obtain the corresponding energy densities for the above relations:

$$\rho_1 = \left[\frac{6\gamma_1 c^2}{4\pi G}\right] R^{-2}, \tag{10.7.13}$$

$$\rho_2 = \frac{\Lambda}{\kappa} = \left[\frac{\gamma_2\gamma_1 c^4}{2\pi G}\right] R^{-2}, \tag{10.7.14}$$

and,

$$\rho_3 = \frac{\left[\frac{L^2}{MR^2}\right]}{\frac{4}{3}\pi R^3} = \left[\frac{3\gamma_1\gamma_3 c^4}{2\pi G}\right] R^{-2}. \tag{10.7.15}$$

We can check that all energy densities are proportional to R^{-2}, so that, we can also write:

$$\rho_{TOT} = \rho_1 + \rho_2 + \rho_3 = \Gamma R^{-2} \qquad (\Gamma = \text{constant}). \tag{10.7.16}$$

In the spirit of inflationary Cosmology (Guth, 1981), we identify, for the present Universe, ρ_{TOT} with the critical density, so that we would have:

$$\rho_{TOT} \cong 2 \text{ x } 10^{-29}\ \text{g / cm}^3.$$

In the next Section, we calculate an estimate for neutrinos mass, and its time variation. But, we observe that, if dark matter is a fraction of ρ_{TOT}, this fraction will also depend on R^{-2}, so as to keep all relative components equally balanced along time.

THEORY FOR NEUTRINOS' ENERGY DENSITY

As we have noticed before the energy density of dark matter, to be identified with neutrinos, shall be given by:

$$\rho_\nu = 0.27\rho_{TOT}. \tag{10.7.17}$$

Berman (2006 d) along with others (see Sabbata and Sivaram, 1994) have estimated that the Universe possess a magnetic field which, for Planck's Universe, was as huge as 10^{55} Gauss. The relic magnetic field of the present Universe is estimated in 10^{-6} Gauss. We can then, suppose that all neutrinos' spins have been aligned with the magnetic field. On the other hand, the spin of the Universe is believed to have increased in accordance with Machian relation (10.7.6) above, which entails relation (10.7.11) above. If we call n the number of neutrinos in the present Universe, and n_{Pl} its value for Planck's Universe, we may write:

$$\frac{n}{n_{Pl}} = \frac{L}{L_{Pl}} = 10^{120}. \tag{10.7.18}$$

Then,

$$n = n_{Pl}\left[\frac{R}{R_{Pl}}\right]^2. \tag{10.7.19}$$

We have just obtained the relation for the increase of the number of neutrinos with R^2. Now, we write the energy density of neutrinos,

$$\rho_\nu \cong \frac{nm_\nu}{\frac{4}{3}\pi R^3}, \tag{10.7.20}$$

where m_ν is the rest mass of the average neutrino.

If we impose relation (10.7.17) and simultaneously, relations (10.7.16) and (10.7.20), we conclude two things:

$$1^{st}.)\quad \rho_\nu = 0.27\rho_{Pl}\left[\frac{R}{R_{Pl}}\right]^{-2}. \tag{10.7.21}$$

$$2^{nd}.)\quad m_\nu = \frac{\rho_{Pl}R_{Pl}^4}{R}. \tag{10.7.22}$$

We see now that while the number of neutrinos in the Universe increases with R^2, the rest mass decreases with R^{-1}; we may obtain, with $R \cong 10^{28}$ cm, that the rest mass of neutrinos should be, in the present Universe:

$$m_\nu \cong 10^{-65}\text{ g}. \tag{10.7.23}$$

Gomide (1963), has estimated the mass of neutrinos a long time ago, finding, in a seminal paper, the value, 10^{-65} g. Gomide(1963) also had equated $m_\nu c^2$ with the self-gravitational energy of the proton.

Conclusions

A law of variation for the number of neutrinos in the Universe has been found. A law of variation for the rest mass of neutrinos was also found.

We remind the reader that Kaluza-Klein's cosmology (Wesson, 1999; 2006; Berman and Som, 1993), consider time varying rest masses, in a penta-dimensional space-time-matter, of which the fifth coordinate is rest mass. The above results can not be rejected, for the time being, by any known data. We point out, that some of the features of the present calculation, are originated from a seminal paper by Sabbata and Gasperini (1979).

The reader must be aware that if $E = 0$, the Universe is also endowed with zero energy density. This is decurrent of the existence of a negative self-gravitacional energy density,

$$\rho_{grav} = \frac{dE_{grav}}{dV} = \frac{1}{4\pi R^2}\frac{dE_{grav}}{dR} = -\frac{1}{4\pi R^2}\left[\frac{2c^4\gamma_1^2}{G}\right]\left[\frac{dR}{dR}\right] = \frac{2c^4\gamma_1^2}{4\pi G}R^{-2}$$

This being the case, relation (10.7.16) does not give the effective energy density of the Machian Universe, which is given in fact by:

$$\rho = \rho_{grav} + \rho_{TOT} = 0$$

The reader should remember that $\rho_{grav} < 0$.

Part V

Concluding Remarks

Chapter 11

Concluding Remarks

11.1. The Present Century's Developments

In order to account for all the interactions in Nature, it would be interesting to superunify gravitation, electromagnetism and particle physics. This can be done by higher-dimensional theories: supersymmetry, supergravity or string theories, for instance. As a first step, it has been put forward a pentadimensional physical theory, whose origin remounts to Kaluza-Klein early attempts in the year 1921 onwards. Today, this theory comprehends three different characteristics: in 5–dimensional extension of the general relativistic framework, there are 15 independent components of the metric tensor: all of them but one, depend on the four spacetime coordinates of GRT; these coefficients are determined by 14 field equations: the ten Einstein's field equations and the four Maxwell's. In this respect, gravitation and electromagnetism are unified. The last component of the metric tensor, represents a scalar field usually attributed to the dark energy's cosmological "constant". There are two different alternative theories with the same aim, "space-time-matter" and "membrane" theories, with a mathematical equivalence. Pentadimensional models, can also account for quantum phenomena, in the low-energy limit of higher-dimensional theories.

In that theory, the four-dimensional sources of matter are absorbed by a 5–dimensional space, which embeds the former one, i.e., the general relativistic theory. In this reduction, the otherwise variable cosmological term, determines the cosmological constant of the original GRT. In the space-time-matter theory, the fifth-dimension is represented by rest-mass. Membrane theory is relatively newer: it is based on a "brane" embedded in the pentadimensional, or higher, space. The counterpart of standard Roberston-Walker's cosmology, is a flat empty 5-D "Universe".

It turns out, that when we apply Mach's Principle into the zero-total-Universe's energy, there is a clear evidence that the Universe behaves as a white-hole (Section 10.6). It can be shown that, from a Machian point of view, the possible angular momentum of Universe is proportional to R_U^2, the radius of the causally connected Universe; the mass M_U, depends linearly on R_U; the cosmological term, with R_U^{-2}; the entropy, like $R_U^{3/2}$ or $M_U^{3/2}$; and the temperature T, like $R_U^{-1/2}$. For a black (or, white)-hole, the entropy has the same mass dependence and, thus, the same temperature dependence ($\propto T^{-3}$); its mass, has the same function as M_U. On the other hand, the spin of the Universe, as it is proportional to R_U^2, is also proportional to M_U^2: it has been verified from astrophysical observations that in such case, the same dependence with the mass, is obtained from astrophysical objects' angular momenta. This is expected in the induced matter theory. For an account of this theory, see Wesson (1999; 2006).

As we have said before, cosmological models in pentadimensional space give rise to GRT's cosmological constant; but, in some cases, they yield a variable "constant". Some pentadimensional models, are Λ – dominated, like in standard 4-D cosmology. A decreasing cosmological "constant", of the type required by Mach's principle, is obtainable in 5-D theories. Of course, other microphysical interactions, are only accommodated in more dimensions.

11.2. Review of the Text

We have seen (Parts I and II) how General Relativity Theory is formulated out of tensor calculus (Chapters 1 and 2) and some basic principles (Chapter 3). A brief chapter includes the Classical Experimental Tests and the Schwarzschild's metric which represent the Solar System or non-rotating static black-holes.

In Part III, we introduced Relativistic Cosmology, and Mach's Principle, which plays an important role in further developments, like the Jordan-Brans-Dicke scalar tensor theories of Gravity. A more detailed approach to all these subjects of Part III may be found in Berman (2007).

The energy of the vacuum and the Λ problem with its possible solution as a dynamical variable, have been addressed in Part IV, Chapters 8 and 9. From a Quantum point of view, this dynamical variable term is currently called *quintessence*. We have avoided this word, however. The zero-energy of the Universe, along with creation out of nothing, was also addressed in those Chapters. An effort has been made, in order to show how the variable lambda solution is compatible with the mathematical and theoretical framework along with the known accelerating present Universe.

In Chapter 10, we further introduce a Jordan-Brans-Dicke cosmological model with variable lambda and or without variable speed of light. Here and then, the Generalized Large Number Hypothesis, is shown to play a significant role in pointing out the directions for variation of several parameters in the Universe, including the possible cases of variable fine structure "constant". Variable rest-mass neutrinos may be responsible for dark matter in the Universe, as in Section 10.7.

We have paved the way for the reader to study up to date papers dealing with the lambda problem with the vantage point of one who became familiar with the basic subject. A reader that is interested on the subjects of this book is supposed to go ahead and consult the various references in the end of this book. In order to orient such reading we suggest, first of all, to read the reviews by V. Sahni. and A. Starobinsky (2000), and T. Padmanabhan (2002). These reviews will certainly complement this book, intended to be read by beginners. A good additional reference on Cosmology and black holes, is Hawking (1993).

In the Appendices, we extend Chapter 4, by further introducing the Lense-Thirring metric (which was verified experimentally in the end of the year 2004), with a treatment of Gravitomagnetism; we briefly mention anti-gravity and energy calculations, including the case of a Kerr-Newman metric. The fate of gravitational collapse is also briefly analyzed.

We also show the most recent discovery with GRT, by Cooperstock and Tieu (2005), which may lead to a dismiss of the concept of "dark" matter, in what refers to galaxies.

We hope that the subjects treated in this book, will remain useful for many years to come. However, it must be pointed out that, ten years ago, if someone would say that the Universe was accelerating and the Einstein–deSitter model would not apply to the present Universe, it might happen that he would be found unable to teach Cosmology. So that, our hopes may not succeed.

Appendix I - Gravitomagnetism

I.1. Introduction

The linearized theory of Gravitation, for a rotating source of matter, shows that there is a precession of a gyroscope in a satellite around such source of matter. In fact, for a satellite around a static spherical distribution, there is a precession. However when the source is rotating, an additional precession term arises. In the last 12 months, two discoveries shook the Physics world: the first was experimental: the Lense-Thirring metric was confirmed by Ignazio Ciufolini and collaborators (Ciufolini, 2005; Ciufolini and Pavlis, 2004). This was a sound evidence in favor of General Relativity Theory. The other discovery was theoretical (Cooperstock and Tieu, 2005), and we shall delve with it in Appendix II bellow: it was Fred Cooperstock and collaborators' discovery that the hypothesis of dark matter is not necessary for understanding the rotation of galaxies according to General Relativity. We now return to the Lense-Thirring metric, to state that within linearized theory, the second precession, a gravitomagnetic effect, is to be studied.

As early as in the year 1974, Papapetrou (1974) found what he called a highly unphysical solution under which, for sufficiently small radial distances, the gravitational force in the Reissner-Nordström metric, is no longer attractive, but repulsive, due to negative energy. Cooperstock and Rosen (1989), referred to it as Reissner-Nordström repulsion and Bonnor and Cooperstock(1989) found the electron´s "mass" as negative, fifteen years later.

When I looked at the three excellent Physical Review D papers written in the year 1990 by K.S. Virbhadra (1990), and the one by Aguirregabiria et al. (1996), I was surprised that the formula he obtained for the energy of a Kerr-Newman black hole, did not reduce to the result published in Adler et al. (1975) for the Schwarzschild´s metric. Even seven years later, in 1996, Aguirregabiria, Chamorro, and Virbhadra (1996) obtained the energy of a Kerr-Newman b.h. for any value of the rotation parameter, with the same lack of reduction to the one in Adler et al. (1975), when the electric charge and the rotation parameter were given zero values. I obtained "correct" values, by employing the Adler et al. pseudotensor formalism (Berman, 2006; 2006a). Furthermore, we identify the possibility of antigravitation production, by means of rapid rotations as described by me,without violation of the Positivity of Energy Theorem. We anticipate that spaceships could use this effect for propulsion purposes.

We study briefly the problem of the end-point in gravitational collapse, suggesting that the infinite point singularity is never attained.

In order to define gravitomagnetism, which is a phenomenon caused by the rotation of a distribution of matter, (likewise, magnetism is caused by rotating electric charges), hence its name, we study first the linearized form of General Relativity metric and field equations, followed by two different derivations of Lense-Thirring metric, that represents the influence of rotation to the otherwise linearization of Schwarzschild's metric. LAGEOS satellites, in orbit around the Earth, measured the Lense-Thirring effect, which is a tiny shift in the orbit of a test-particle(Ciufolini, 2005; Ciufolini and Pavlis, 2004). Frame-dragging, gravito-magnetic field and Lense-Thirring effect, are predictions of Einstein's theory of General Relativity. The axes of a freely falling frame, where the equivalence principle holds, are not fixed relative to, say, the distant "fixed stars", but they are dragged by any moving mass, including the rotation of the Earth: this is frame-dragging or the "dragging of inertial frames", according to Einstein. The orbit of a test-particle, as a small satellite, constitutes an experimental gyroscope, with nodal and pericenter perturbations. The LAGEOS satellites measured the frame-dragging of the Earth to be 99% of the value predicted by General Relativity.

Gravity Probe B experiment will try to measure gravitomagnetism generated by the rotation of the Earth on a gyroscope, while the WEBER-SAT should measure the Earth angular-momentum effect on the orbiting satellite. In some non-metric theories of gravity, the two effects are different; this is not the case with General Relativity Theory.

I.2. Linearized General Relativity

We review the linearized form of General Relativity, which theory, becomes Lorentz-covariant. The metric is assumed as quasi-Minkowskian(D'Inverno, 1992):

$$g_{\mu\nu} = \eta_{\mu\nu} + h_{\mu\nu}, \tag{AI.2.1}$$

with:

$$|g_{\mu\nu}| \gg |h_{\mu\nu}|,$$

and,

$$\lim_{r \to \infty} |h_{\mu\nu}| = 0,$$

so that,

$$g^{\mu\nu} = \eta^{\mu\nu} - h^{\mu\nu} \quad . \tag{AI.2.2}$$

Indices are raised and lowered with $\eta^{\mu\nu}$ and $\eta_{\mu\nu}$ respectively. The Ricci tensor is, now given by:

$$R_{\alpha\beta} = \eta^{\gamma\delta} R_{\gamma\alpha\delta\beta} = \frac{1}{2}\left[h^{\gamma}_{\alpha,\beta\gamma} + h^{\gamma}_{\beta,\alpha\gamma} - \Box^2 h_{\alpha\beta} - h_{,\alpha\beta}\right], \tag{AI.2.3}$$

where

$$h \equiv h^{\gamma}_{\gamma}, \tag{AI.2.4}$$

and,

$$\Box^2 \equiv \frac{\partial^2}{\partial t^2} - \nabla^2. \qquad \text{(AI.2.5)}$$

$$\Box^2 \text{ and } \nabla^2$$

stand respectively for the d'Alembertian and 3-dimensional Laplacian in Minkowski's space. The curvature scalar is, now:

$$R = h^{\gamma\delta}_{,\gamma\delta} - \Box^2 h, \qquad \text{(AI.2.6)}$$

With appropriate gauge transformations, the field equations can be put in the following form:

$$\Box^2 \left(h_{\mu\nu} - \tfrac{1}{2} h \eta_{\mu\nu}\right) = 2\kappa T_{\mu\nu}, \qquad \text{(AI.2.7)}$$

with

$$h^{\mu\nu}_{,\nu} = 0. \qquad \text{(AI.2.8)}$$

We are supposing that the space coordinates are isotropic, so that:

$$h_{11} = h_{22} = h_{33} = h_s. \qquad \text{(AI.2.9)}$$

Furthermore, we suppose that the matter source is slow-moving and constituted by low density matter with a dust equation of state: $p = 0$. In this case, the perfect fluid energy-momentum tensor is represented by:

$$T^{\mu\nu} = \rho u^\mu u^\nu, \qquad \text{(AI.2.10)}$$

where, u^μ is the quadrivelocity and ρ represents the matter density. The perturbation is considered in its lower order.

With these considerations, we now may tackle the derivation of Lense-Thirring metric, in next section.

I.3. Lense-Thirring Metric Derivation

We shall derive Lense-Thirring metric by two different lines of argument. The first will be by means of linearized General Relativity, outlined in the last section. The other is a transformation argument apparently first presented in the reference by Adler and Silbergleit (2000), which is the basis for this Appendix I.

Referring to the linearized field equations derived in last section, namely:

$$\Box^2 \left(h_{\mu\nu} - \tfrac{1}{2} h \eta_{\mu\nu}\right) = 2\kappa T_{\mu\nu}, \qquad \text{(AI.3.1)}$$

and the gauge condition stated earlier:

$$h^{\mu\nu}_{,\nu} = 0, \tag{AI.3.2}$$

which is the analog of the same condition in electromagnetism, we have:

$$h_{00} = h_s = 2\Phi. \tag{AI.3.3}$$

In Schwarzschild's metric we would have:

$$\Phi_{schw} = -\frac{GM}{2r}. \tag{AI.3.4}$$

However, here the field equations are different:

$$\Box^2\Phi = -\kappa\rho, \tag{AI.3.5}$$

$$\Box^2\vec{h} = 2\kappa\rho\vec{v}, \tag{AI.3.6}$$

where $\vec{h} \equiv (h_{01}, h_{02}, h_{03})$ and $\vec{v}$ represents the source velocity. (In the same token, we shall obtain a different metric than Schwarzschild's).

We know from electromagnetism, how to solve (AI.3.5) and (AI.3.6) by means of retarded time potentials. The equation (AI.3.5) yields a scalar function $\Phi(\vec{r}, t)$ which is called gravitoelectric potential. The second equation, (AI.3.6), yields a new phenomenon called gravitomagnetism, represented by the gravitomagnetic vector potential $\vec{h}(\vec{r}, t)$.

When the system is nearly time-independent, the retarded time effect can be neglected, and we find, a Newtonian potential (gravitoelectric-field) and a relativistic vector potential (gravitomagnetic field):

$$\Phi(\vec{r}) = -G\int \frac{\rho(\vec{r}\,')}{|\vec{r}-\vec{r}\,'|} d^3\vec{r}\,', \tag{AI.3.7}$$

and,

$$\vec{h}(\vec{r}) = 4G\int \frac{\rho(\vec{r}\,')\,\vec{v}(\vec{r}\,')}{|\vec{r}-\vec{r}\,'|} d^3\vec{r}\,'. \tag{AI.3.8}$$

We have, thus, obtained the metric ("Lense-Thirring"):

$$ds^2 = (1+2\,\Phi)dt^2 - (1-2\,\Phi)d\vec{r}^{\,2} + 2\vec{h}\cdot d\vec{r}\,dt. \tag{AI.3.9}$$

The last term in r.h.s., accounts for gyroscope precessions ("Lense-Thirring effect").

I.4. Alternative Derivation for the Lense-Thirring Metric

We now shall consider the isotropic Schwarzschild's solution (Adler et al., 1975):

$$ds^2 = \frac{\left[1-\frac{m}{2r}\right]^2}{\left[1+\frac{m}{2r}\right]^2} dt^2 - \left[1-\frac{m}{2r}\right]^4 d\vec{r}^{\,2}, \tag{AI.4.1}$$

where we expand the metric coefficients, supposing that $\frac{m}{r} \ll 1$, finding:

$$ds^2 = \left(1 - \tfrac{2m}{r} + \tfrac{2m^2}{r^2} + \cdots\right) dt^2 - \left(1 + \tfrac{2m}{r} + \tfrac{3m^2}{r^2} + \cdots\right) d\,\vec{r}^{\,2}. \qquad \text{(AI.4.2)}$$

Neglecting quadratic and higher order terms, we obtain:

$$ds^2 \cong \left(1 - \tfrac{2m}{r}\right) dt^2 - \left(1 + \tfrac{2m}{r}\right) d\,\vec{r}^{\,2}. \qquad \text{(AI.4.3)}$$

As we have a nearly Lorentzian metric, we may transform to a moving source coordinate system, by means of a linearized Lorentz transformation, which, for low speeds and being v the velocity in the, say, x−direction, i.e.,

$$t_r = t - vx, \qquad \text{(AI.4.4)}$$

$$x_r = x - vt. \qquad \text{(AI.4.5)}$$

In the above, the index "r" stands for rest system time or space coordinates.

The moving source metric, becomes now:

$$ds^2 = \left(1 - \tfrac{2m}{r}\right) dt^2 - \left(1 + \tfrac{2m}{r}\right) d\,\vec{r}^{\,2} + \tfrac{4m}{r} v\, dx\, dt. \qquad \text{(AI.4.6)}$$

For motion in any direction, relation (AI.4.6) becomes:

$$ds^2 = \left(1 - \tfrac{2m}{r}\right) dt^2 - \left(1 + \tfrac{2m}{r}\right) d\,\vec{r}^{\,2} + \tfrac{4m}{r} \left(\vec{v} \cdot d\vec{r}\right) dt. \qquad \text{(AI.4.7)}$$

The above metric represents a point-mass source. We may superpose an infinite number of such point masses, generating a whole continuum distribution source, and then,

$$\tfrac{m}{r} \longrightarrow \Phi(\vec{r}),$$

and

$$\tfrac{4m}{r}\vec{v} \longrightarrow \vec{h}(\vec{r}),$$

where $\Phi(\vec{r})$ and $\vec{h}(\vec{r})$ are defined by relations (AI.4.7) and (AI.4.8) of last section.

I.5. Precession of Gyroscopes. Contact with Lageos Experiment

As we have already mentioned, in the end of the year 2004, experimental evidence showed that gyroscopes in orbit around the Earth, precessed according to Lense-Thirring metric predictions. We shall now introduce the main details to understand such precession.

Consider the energy-momentum tensor $T_{\mu\nu}$ of a closed system, which is then conserved:

$$T^{\mu\nu}{}_{;\nu} = 0.$$

In Special Relativity, the corresponding equation is:

$$\tfrac{\partial}{\partial x^\nu} T^{\mu\nu} = 0.$$

We now construct the tensor $M^{\alpha\beta\gamma}$:

$$M^{\gamma\alpha\beta} \equiv x^{\alpha} T^{\beta\gamma} - x^{\beta} T^{\alpha\gamma}.$$

If $T^{\mu\nu} = T^{\nu\mu}$, $M^{\gamma\alpha\beta}$ is also conserved. We may now define a "total" angular momentum,

$$J^{\alpha\beta} = \int d^3x \, M^{0\alpha\beta} = -J^{\beta\alpha}.$$

By adequately fixing the origin of our coordinate system, we may make $J^{0i} = 0$ $(i = 1,2,3)$. Then, J^{23}, J^{31} and J^{12}, are the components of the angular momentum, in the usual sense. As in Newtonian gravity, we may obtain the intrinsic spin, by isolating the internal part of $J^{\alpha\beta}$; in that case the spin is a constant. In order to do that, we make a translation of coordinate system:

$$x^{\alpha} \to x'^{\alpha} = x^{\alpha} + a^{\alpha}; \text{ we find,}$$

$$J^{\alpha\beta} \to J'^{\alpha\beta} = J^{\alpha\beta} + a^{\alpha} p^{\beta} - a^{\beta} p^{\alpha}.$$

We recognize p^{α} as the linear momentum, and we define the spin-vector S_{α} by means of:

$$S_{\alpha} \equiv \tfrac{1}{2} \varepsilon_{\alpha\beta\gamma\delta} J^{\beta} U^{\delta}.$$

In the above, $U^{\alpha} = \frac{dx^{\alpha}}{ds}$ (quadrivelocity), and $\varepsilon_{\alpha\beta\gamma\delta}$ is the well-known completely anti-symmetric unitary tensor.

For a free particle, then:

$$\frac{dS_{\alpha}}{dt} = 0.$$

In the center-of-mass system:

$$U^i = 0,$$

$$U^0 = 1,$$

$$S_1 = J^{23},$$

$$S_2 = J^{31},$$

$$S_3 = J^{12},$$

$$S_0 = 0.$$

We would have, in this coordinate system:

$$U^{\alpha} S_{\alpha} = 0.$$

The same above relation, translates into a general coordinate system, unchanged. In General Relativity the corresponding relation $\frac{dS_\alpha}{dt} = 0$ would translate to:

$$\frac{DS_\alpha}{Ds} = 0,$$

or, if we follow Adler and Silbergleit (2000), the parallel displacement equation for the "gyro" spin vector is:

$$\frac{dS^\alpha}{ds} + \left\{ \begin{matrix} \alpha \\ \nu\ \beta \end{matrix} \right\} S^\nu \frac{dx^\beta}{ds} = 0.$$

We shall work with the first order approximation, in the velocities $\vec{v}$ and $\vec{V}$, respectively for the Earth and the gyroscope. We suppose that S^α is orthogonal to the quadrivelocity, i.e., $S^0 = 0$, in the rest frame. In another frame, in this order of approximation,

$$S^0 = \vec{S} \cdot \vec{V}$$

In the lowest order, for the Lense-Thirring metric, the corresponding Christoffel symbols would be:

$$\left\{ \begin{matrix} 0 \\ 0\ i \end{matrix} \right\} = \Phi_{,i},$$

$$\left\{ \begin{matrix} j \\ 0\ 0 \end{matrix} \right\} = \Phi_{,j},$$

$$\left\{ \begin{matrix} j \\ j\ j \end{matrix} \right\} = \Phi_{,j},$$

$$\left\{ \begin{matrix} j \\ j\ i \end{matrix} \right\} = -\Phi_{,i},$$

$$\left\{ \begin{matrix} i \\ j\ j \end{matrix} \right\} = \Phi_{,i},$$

$$\left\{ \begin{matrix} i \\ 0\ j \end{matrix} \right\} = \tfrac{1}{2}(h_{j,i} - h_{i,j}).$$

In the above, $i, j = 1, 2, 3$, and $i \neq j$.

The gravitomagnetic (vector) potential $\vec{h}$ is reduced, by its gauge transformation, so that it appears only under its "curl".

The spin equation, now reads:

$$\dot{S}^i + 2\Phi_{,i}\left(S^j V^i\right) - S^i \Phi_{,j} V^j - V^i \Phi_{,j} S^j + \tfrac{1}{2} S^j \left(h_{j,i} - h_{i,j}\right) = 0.$$

The scalar potential Φ gives raises to a "geodetic" precession term $\vec{\Omega}_G$, while the smaller Lense-Thirring vector potential $\vec{h}$ yields a precession rate $\vec{\Omega}_{LT}$.

It was the Lense-Thirring effect, that was verified in the year 2004. This is the first verification of GRT's gravitomagnetism, outside Schwarzschild's metric, within the Solar system

I.6. Kerr-Newman Gravitomagnetism

By means of energetic considerations, we find a general relation for gravitomagnetic energy associated with Kerr-Newman metric.

We recall the derivation of Lense-Thirring metric as above: the most general black hole is characterized by the "exact" rotating metric with mass M, electric charge Q and rotational parameter "a" and is given by Kerr-Newman's metric, where in quasi-Cartesian form, is given by (Newman et al., 1965):

$$ds^2 = dt^2 - dx^2 - dy^2 - dz^2 - \frac{2\left[M - \frac{Q^2}{2r_0}\right] r_0^3}{r_0^4 + a^2 z^2} \cdot F^2 \tag{AI.6.1}$$

$$F = dt + \frac{Z}{r_0} dz + \frac{r_0}{\left(r_0^2 + a^2\right)} (xdx + ydy) + \frac{a(xdy - ydx)}{a^2 + r_0^2} \tag{AI.6.2}$$

$$r_0^4 - \left(r^2 - a^2\right) r_0^2 - a^2 z^2 = 0 \tag{AI.6.3}$$

and

$$r^2 \equiv x^2 + y^2 + z^2 \tag{AI.6.4}$$

In order to compare the Lense-Thirring metric with Kerr-Newman's, we consider the latter, in the equivalent Boyer-Lindquist(Boyer and Lindquist, 1967) coordinates, in the $Q = 0$ case, which can be cast as:

$$ds^2 = \left[\frac{r^2 + a^2\cos^2\theta - 2mr}{r^2 + a^2\cos^2\theta}\right] dt^2 + \frac{4mar\sin^2\theta d\phi dt}{r^2 + a^2\cos^2\theta} - \left\{\frac{[(r^2+a^2)(r^2+a^2\cos^2\theta)] + [2mra^2\sin^2\theta]}{r^2 + a^2\cos^2\theta}\right\} \sin^2\theta d\phi^2 -$$

$$-\left(r^2 + a^2\cos^2\theta\right)\left(d\theta^2 + \frac{dr^2}{r^2 + a^2 - 2mr}\right) \quad ,$$

where, when we compare with Lense-Thirring metric, valid in the long-range case ($a << 1$), we find:

$$m \equiv M,$$

and,

$$aM = \text{angular momentum in the } \theta = 0 \text{ direction.}$$

It can be checked that the introduction of a $Q \neq 0$ electric charge, was done in such a way that the metric should reduce, in the non-rotating case ($a = 0$), to the well-known Reissner-Nordström's metric, which represents the generalization of Schwarzschild's metric with an electric charge. Though, we have not proved the Kerr-Newman's metric, the above reductions, render it very plausible.

After a lengthy calculation, we find, for the Kerr-Newman's metric (in $G = c = 1$ units)(Berman, 2006; 2006a):

$$P_0 = M - \left[\frac{Q^2+M^2}{4\rho}\right]\left[1+\frac{(a^2+\rho^2)}{a\rho}arctgh\left(\frac{a}{\rho}\right)\right] \tag{AI.6.9}$$

$$P_1 = P_2 = P_3 = 0 \tag{AI.6.10}$$

The last result "validates" the coordinate system chosen for the present calculation: it is tantamount to the choice of a center-of-mass coordinate system in Newtonian Physics, or the use of comoving observers in Cosmology.

By considering an expansion of the arcth($\frac{a}{\rho}$) function, in terms of increasing powers of the parameter $\left(\frac{a}{\rho}\right)$, and by neglecting terms of the third or larger powers, we find the energy of a slowly rotating Kerr-Newman black-hole,

$$E \simeq M - \left[\frac{Q^2+M^2}{R}\right]\left[\frac{a^2}{3R^2}+\frac{1}{2}\right] \tag{AI.6.11}$$

where $\rho \to R$; this can be seen because the defining equation for ρ is:

$$\frac{x^2+y^2}{\rho^2+a^2}+\frac{z^2}{\rho^2} = 1 \text{ and if } a \to 0, \rho \to R.$$

We can interpret the terms $\frac{Q^2a^2}{3R^3}$ and $\frac{M^2a^2}{3R^3}$ as the magnetic and gravitomagnetic energies caused by rotation. Virbhadra(1990) and Aguirregabiria et al. (1996) noticed the first of these effects in the year 1990, but since then it seems that they failed to recognize the existence of the gravitomagnetic energy due to M, on an equal footing.

Furthermore, we can, from relation (AI.6.11), find the energy density associated with the black-hole:

$$\mu = \frac{1}{4\pi R^2}\frac{dE}{dR} = \frac{1}{4\pi R^4}\left[Q^2+M^2\right]\left[\frac{a^2}{R^2}+\frac{1}{2}\right] \tag{AI.6.12}$$

As expected, the energy density is essentially positive, so it obeys the weak energy condition. The gravitomagnetic component of the energy density is, given by $\left[\frac{M^2a^2}{4\pi R^6}\right]$.

We shall now show how to attain a "repulsive" gravitation (or antigravitation) state identified with negative energy contents in a limited region of space, without violation of the positivity of energy.

The total energy E of an isolated system should be positive in General Relativity, in accordance with the "Positive Energy Theorem", of Schoen and Yau, Choquet-Bruhat, Deser, Teitelboim, Witten, York, etc, as cited by Ciufolini and Wheeler(1995). The exception is Minkowski's space, where $E = 0$. (See also Weinberg, 1972). The condition usually stated for these theorems is that the dominance of energy condition be satisfied, which entails the weak energy condition $\mu \geqslant 0$, where μ stands for the energy density (Ciufolini and Wheeler,1995), while the Einstein field equations are taken as valid.

Even in popular books, like Hawking's best-seller (Hawking, 2001), it is stated that (total) energy in General Relativity is positive. The rule is broken by the energy of the Universe, $E = 0$, as stated in Hawking's book (Hawking, 2001), because zero is not positive, and the Universe's metric is not Minkowskian (see Berman,2006).

Ciufolini and Wheeler (1995) also comment that an electric charge circling in orbit creates magnetism and, likewise, a spinning mass creates gravitomagnetism; the gravitomagnetic effect due to the slow rotation of the Earth is measurable, by an orbiting gyroscope. (Lageos III; Gravity Probe B, etc.). However, we are far from a spinning black-hole (Ciufolini and Wheeler,1995).

When obtaining the central mass field, or Kerr metric, for a spinning massive object, one has to measure the energy contents of the distribution, encircled by a radial distance R, and we have shown that, for non-relativistic (slow) rotations, the energy contents is given by (within a radial distance R):

$$E = Mc^2 - \frac{GM^2}{2R} - \frac{a^2M^2}{3R^3} \tag{AI.6.13}$$

where M, c G, R and "a", stand respectively for central mass, speed of light in vacuum, gravitational constant, radial distance, and angular momentum per unit mass, and $Q = 0$.

The first rhs term above is the inertial energy, as known from Special Relativity; the second, is the relativistic equivalent to the self-gravitational energy, (in Newton's theory $-\frac{3}{5}\left(\frac{GM^2}{R}\right)$).

The 3rd., stands for gravitomagnetic energy.

A similar calculation, employing, Berman's suggestions, show that the angular momenta of the Kerr-Newman black hole, includes a gravitomagnetic contribution, in the slow rotation case: Berman(2006j)

$$\Delta J \cong -2M^2 \left[\frac{a^3}{5R^3} + \frac{a}{3R}\right].$$

An embarrassing problem with (AI.6.13), is that when R becomes, for instance, $R_0 \leqslant \frac{GM}{2c^2}$, we have $E < 0$. This does not violate the positivity of "total" energy theorem, but we interpret as pertaining to a negative energy that represents a repulsive gravity region (anti-gravitation).

This effect could be used to motorize an antigravitational engine. The technological effort necessary to attain this objective may only succeed at about 20 years from now. We refer to the above equation for E in the rapidly spinning case, with Kerr metric ($Q = 0$),

$$E = Mc^2 - \frac{GM^2}{4\rho}\left[1 + \left(\frac{a^2+\rho^2}{a\rho}\right) arctgh\left(\frac{a}{\rho}\right)\right] \tag{AI.6.14}$$

where $\frac{x^2+y^2}{\rho^2+a^2} + \frac{z^2}{\rho^2} = 1$ and $\rho \geqslant 0$, and E stands for the energy inside a surface of constant ρ values.

In fact, if we interpret the total energy, as the limit, of E, when the radial distance goes to infinity, so that the "entire" space is involved (say, put observer at spatial infinity), the "positivity" of total energy will be preserved, for a positive mass M, while we would still have, at certain "radial" distances ρ = constant, the possibility for E < 0 (antigravitation or repulsive gravity). We must, indeed, not confuse the "total" energy (over all space), with the energy distribution contents at radial distance from the source of the field. The "total" energy is, of course, $\lim_{R\to\infty} E = Mc^2 > 0$. We conclude that antigravity can be obtained in practice, but the total energy of a black hole is the product of the "mass parameter" M with the square of the speed of light c^2.

When a certain charged rotating body enters in gravitational collapse, a Kerr-Newman black-hole is formed(Newman et al., 1965). Only three parameters define this "hole": its mass M, charge Q, and rotational parameter "a", such that $a = -\frac{J}{M}$, where J is its total angular-momentum. The energy contents, within a "radial" distance ρ, of a Kerr-Newman b.h., may be found from the relation derived in Section AI.1, with $Q = 0$, to wit,

$$E = Mc^2 - \frac{GM^2}{4\rho}\left[1 + \left(\frac{a^2+\rho^2}{a\rho}\right) arctgh\left(\frac{a}{\rho}\right)\right] \tag{AI.6.15}$$

where ρ is the positive root of:

$$\frac{x^2+y^2}{\rho^2+a^2} + \frac{z^2}{\rho^2} = 1. \tag{AI.6.16}$$

While collapse is in process, ρ will become smaller, up to the point when E becomes null, and, afterwards, negative. We have suggested that when $E \leq 0$, the negative energy contents E within a certain ρ value, is characteristic of repulsive gravitation, or antigravity.

In such case, the collapse will be decelerated, and eventually, halted at a non-null final radius, and then reversed towards a radially increasing direction: the outcome is a surging white-hole, originated from a non-null initial radius, which is also the mentioned end-point of gravitational collapse.

The endpoint of gravitational collapse may be, thus, NOT a black-hole with singular point of infinite energy density, but an expanding white-hole(Berman, 2004 a). The positivity of total energy, is still given by Mc^2, which is positive for ordinary matter. We can get a taste of involved energy densities in the halting point, by obtaining the energy inside a spherical non-rotating "ball", with $Q = a = 0$, where,

$$E = Mc^2 - \frac{GM^2}{2R} \quad . \tag{AI.6.17}$$

We obtain for the energy density μ:

$$\mu = \frac{dE}{dV} = \frac{1}{4\pi R^2}\frac{dE}{dR} = \frac{GM^2}{8\pi R^4}. \tag{AI.6.18}$$

For $E = E_0 = 0$ we find:

$$R = R_0 = \frac{GM}{2c^2}, \tag{AI.6.19}$$

while:

$$\mu = \mu_0 = \frac{2}{\pi} G^{-3} M^{-2} c^8. \tag{AI.6.20}$$

The mass density is obtained by dividing μ by c^2, so that, for the Sun,

$$\mu_{0Sun}/c^2 \approx 10^{21} \qquad \text{g/cm}^3.$$

The general result is, numerically, when M is given in grams,

$$\mu_0/c^2 \approx 10^{87} M^{-2} \quad \text{g/cm}^3.$$

For the whole Universe, we estimate the very early mass as $\approx 10^{-5}$ grams, which is Planck's mass, so that the maximum big-crunch mass density would be $\approx 10^{97}$ g/cm^3.

I.7 - Conclusion

After reviewing the basis for establishing gravitomagnetism in General Relativity, we cite the "correct" formula for the energy of Kerr-Newman black holes. Though it was obtained by means of Adler et al.'s pseudotensor formalism,it points in the same direction of other authors, like Donald Lynden-Bell, A. Ori and J. Katz (Lynden-Bell, 1985); (Katz and Ori, 1990). It renders the results of Virbhadra (Virbhadra,1990); (Aguirregabiria et al., 1996) difficult to accept, in terms of physical arguments, and by the non-reduction to the calculation in Adler et al. (1975). We augment the notion of repulsive gravity or antigravity beyond the limit of Reissner-Nordström case, and predict spaceships moved by engines employing this effect. We find a gravitomagnetic energy effect, while studying Kerr's metric, as a particular case, and, of course, gravitomagnetism is also a feature in Kerr-Newman's case. We prefer to say that the "mass" of the electron is positive indeed; it is not to be confused with the energy contents inside its "radius", which may be negative.

We have also hinted that the end-point of gravitational collapse includes a non-null finite radius, and may not be the infinite point singularity. This was, indeed, the point of view defended by Albert Einstein, who died before the term "black-hole" had been coined. If the Universe is to be regarded, like we presented here, as a black-hole, the end-point of the "big-crunch" will not be the infinite singularity, but the inception of a new expansion. By the same token, the reversed phenomenon ("big-bang"), would not begin with an infinite singularity, too, but should have an initial non-zero radius; loosely, we may identify the initial radius with Planck's length.

Appendix II - Absence of "Dark" Matter in Galaxies

"Dark" matter, is a conception originally based on galactic rotation observations, which seem to show, within Newtonian theory, that all spirals might have a spherical distribution of non-visible mass. As the angular velocities seem to be constant, even at far distances from the galactic center, it was believed, before the year 2005, that galaxies behave like a rotating solid sphere, where tangential speeds increased linearly with distance from the galactic center. The visible material in the galaxy's plane, would then be a minor fraction of the "total" mass.

In the year 2005, Cooperstock and Tieu(2005), calculated, by means of GRT, that NO "dark" matter is necessary in order to explain the tangential speeds for rotating matter. They modeled a galaxy as a stationary axially symmetric pressure-free fluid, in the weak-field approximation, obtaining, from GRT field equations, for a metric of the type:

$$ds^2 = -e^{\nu-\omega}(udz + dr^2) - r^2 e^{-\omega} d\phi^2 + e^{\omega}(cdt - Nd\phi)^2, \tag{AII.1}$$

where u, ν, ω, and N are functions of the cylindrical polar coordinates (r,z). To the order of approximation required, they made $u \cong 1$. It should be noted that this model is different from the Schwarzschild's solution, because in the latter, the orbiting stars, would be represented by infinitesimal particles moving in the far exterior of the central mass, with no masses in-between.

They chose a comoving frame, with quadrivelocities as in usual Cosmology,

$$U^i = \delta^i_0. \tag{AII.2}$$

With (r,z) fixed, they performed the transformation:

$$\phi \rightarrow \phi + \omega(r,z)t, \tag{AII.3}$$

which diagonalizes the metric locally.

Then, the (local) angular velocity ω, and the tangential speed V, are related as:

$$V \cong \omega r, \tag{AII.4}$$

$$\omega \cong \frac{Nce^{\omega}}{r^2 e^{-\omega} - N^2 e^{\omega}} \cong \frac{Nc}{r^2}. \tag{AII.5}$$

From Einstein's field equations, they found:

$$2r\frac{\partial\nu}{\partial r}+\left(\frac{\partial N}{\partial r}\right)^2-\left(\frac{\partial N}{\partial z}\right)^2=0, \quad \text{(AII.6)}$$

$$r\frac{\partial\nu}{\partial z}+\frac{\partial N}{\partial r}-\frac{\partial N}{\partial z}=0, \quad \text{(AII.7)}$$

$$\left(\frac{\partial N}{\partial r}\right)^2+\left(\frac{\partial N}{\partial z}\right)^2+2r^2\left(\frac{\partial^2\nu}{\partial r^2}+\frac{\partial^2\nu}{\partial z^2}\right), \quad \text{(AII.8)}$$

$$\frac{\partial^2 N}{\partial r^2}+\frac{\partial^2 N}{\partial z^2}=\frac{1}{r}\frac{\partial N}{\partial r}, \quad \text{(AII.9)}$$

$$\frac{8\pi G\rho}{c^2}=\left[\frac{\partial^2\omega}{\partial r^2}+\frac{\partial^2\omega}{\partial z^2}+\frac{1}{r}\frac{\partial\omega}{\partial r}\right]+\frac{3}{4}r^{-2}\left(\frac{\partial N}{\partial r}\right)^2+\left(\frac{\partial N}{\partial z}\right)^2+$$

$$+Nr^{-2}\left[\frac{\partial^2 N}{\partial r^2}+\frac{\partial^2 N}{\partial z^2}-\frac{1}{r}\frac{\partial N}{\partial r}\right]-\frac{1}{2}\left[\frac{\partial^2\nu}{\partial r^2}+\frac{\partial^2\nu}{\partial z^2}\right]. \quad \text{(AII.10)}$$

The geodesic equation, which is the equation of motion, is given by:

$$\frac{dU^i}{ds}+\Gamma^i_{kl}U^kU^l=0. \quad \text{(AII.11)}$$

For a comoving observer, then, it follows that (within the fluid):

$$\frac{\partial\omega}{\partial r}=\frac{\partial\omega}{\partial z}=0, \quad \text{(AII.12)}$$

so that:

$$\nabla^2\omega=\frac{\partial^2\omega}{\partial r^2}+\frac{\partial^2\omega}{\partial z^2}+\frac{1}{r}\frac{\partial\omega}{\partial r}=0. \quad \text{(AII.13)}$$

From the field equations,

$$\frac{\partial^2 N}{\partial r^2}+\frac{\partial^2 N}{\partial z^2}-\frac{1}{r}\frac{\partial N}{\partial r}=0. \quad \text{(AII.14)}$$

$$r^{-2}\left(\frac{\partial N}{\partial r}\right)^2+\left(\frac{\partial N}{\partial z}\right)^2=\frac{8\pi G\rho}{c^2}. \quad \text{(AII.15)}$$

Cooperstock and Tieu state that N is of order $G^{1/2}$, and as the equation relating N and ρ is non-linear, the Newtonian calculations are not valid.

From equation (AII.14), we find that:

$$\nabla^2\left[\int\frac{N}{r}dr\right]=0. \quad \text{(AII.16)}$$

Then,

$$\nabla^2\Phi=0, \quad \text{(AII.17)}$$

where,

$$\Phi=\int\frac{N}{r}dr. \quad \text{(AII.18)}$$

Remembering (AII.4) and (AII.5), we find, for the tangential speed,

$$V = c\frac{\partial\Phi}{\partial r}. \tag{AII.19}$$

The Laplace equation for Φ, in cylindrical polar coordinates, leads to the solution below, by separation of variables:

$$\Phi = Ce^{-k|z|}J_0(kr). \tag{AII.20}$$

where J_0 stands for the Bessel function of zeroth order, and C = constant.

The general solution is thus, a superposition of the above solution, namely:

$$\Phi = \sum_n C_n e^{-k_n|z|}J_0(k_n r). \tag{AII.21}$$

Then,

$$V = c\frac{\partial\Phi}{\partial r} = c\sum_n C_n k_n e^{-k_n|z|}J_1(k_n r). \tag{AII.22}$$

For the Milky Way, the same authors found that the first ten functions C_i, $(i = 1, ..., 10)$, were an enough approximation. You compute Φ first, then you calculate N, and find then ρ; you compare with observed data, to check whether the approach is correct.

Now let us go to the resultant picture for the Milky Way, with similar results for other galaxies:

1) $M \cong 21 \cdot 10^{10} M_\odot$. This result puts the Milky Way mass, at the lowest previous estimates for its mass.

2) The mass distribution, is a flattened disk, and NOT a spherical one.

3) The flat velocity curve is achieved with a disk mass one order of magnitude smaller than the expected hypothetical exotic dark mass magnitude so there is no significant halo-mass (exotic dark matter).

4) Any substantial amount of non visible matter would have to lie in the flattened region at $z = 0$ neighborhood (i.e., in the region of appreciable ρ values); however this ρ value is constituted of normal baryonic matter debris.

5) Conventional GRT accounts for the observed visible mass distribution and rotating speeds. This does not preclude the existence of other dark matter sources in the Universe, like, for instance, massive neutrinos (Section 10.7).

The astonishing discovery of Cooperstock and Tieu was a subject matter in the well-known magazine "The Economist", and is a major break-through discovery at very beginning of the $21^{st.}$ Century. It may well point out to a Nobel prize award for the same authors!

References and Suggested Bibliography

Before any other references (Part RIII), we shall list John D. Barrow's publications (Part RI), and his books (Part RII), which we rank as the best set of references on General Relativity and Cosmology, from a single author, alive or not.

Part RI - Publications by John D. Barrow

[1] Light Elements and the Isotropy of the Universe, 1976. *Mon. Not. Roy. astr. Soc.*, **175**, 359-370.

[2] The Synthesis of Light Elements in Turbulent Cosmologies, 1977. *Mon. Not. Roy. astr. Soc.*, **178**, 625-650.

[3] A Chaotic Cosmology, 1977. *Nature*, **267**, 117-120.

[4] On the Origin of Cosmic Turbulence, 1977. *Mon. Not. Roy. astr. Soc.*, **179**, 47P-49P.

[5] The Homogeneity and Isotropy of the Universe, 1977. *Mon. Not. Roy. astr. Soc.,* **181**, 719-728 (with R A Matzner).

[6] Modern Cosmological Models, 1978. *Science Progress* **65**, 129-160.

[7] Quiescent Cosmology, 1978. *Nature*, **272**, 211-215.

[8] Primordial Black Hole Formation in an Anisotropic Universe, 1978. *Mon. Not. Roy. astr. Soc.,* **182**, 537-558 (with B J Carr).

[9] A Cosmological Limit on the Possible Variation of G, 1978. *Mon. Not. Roy. astr. Soc.,* **184**, 677-682.

[10] Eternity is Unstable, 1978. *Nature*, **276**, 453-459 (with F J Tipler).

[11] Analysis of the Singularity Studies of Belinskii, Lifshitz and Khalatnikov, 1979. *Phys. Reports,* **56**, 371-402 (with F J Tipler).

[12] Pre-galactic Black Holes: A New Constraint, 1979. *Gen. Rel. Gravn.,* **11**, 633-638 (with J Silk).

[13] The Structure of the Early Universe, 1980. *Scientific American*, April issue, pp118-129 (with J Silk).

[14] The Early Stages of the Universe, 1980. Chapter 4, *Proc. of the General Meeting of the European Physical Society*, Inst. of Physics, Heinemann, London.

[15] Galaxy Formation: The First Million Years, 1980. *Phil. Trans. Roy. Soc.* **A296**, 273-288.

[16] Black Holes are not the end of the World, 1979. P*hys. Bulletin*, **30**, 149-150.

[17] Baryon Asymmetry and Cosmological Asymmetries, 1979. *Mon. Not. Roy. astr. Soc.*, **188**, 23P-29P.

[18] Evolutionary Cosmological Models, 1979. In 'Problems of the Cosmos', (Einstein Centenary Volume, Enciclopedia Italiana, Rome).

[19] Shear Hell-Holes and Anisotropic Universes, 1979. *Gen. Rel. Gravn.*, **11**, 383-389 (with B J Carr).

[20] The Size of a Bouncing Mixmaster Universe, 1980. *Phys. Rev. D* **21**, 336-340 (with R A Matzner).

[21] The Baryon Asymmetry of the Universe, 1980. *Surveys in High Energy Physics,* **1**, 1-30.

[22] The Proton Half-life and the Dirac Hypothesis, 1979. *Nature*, **282**, 698-9.

[23] Primordial Baryon Generation by Black Holes, 1979. *Mon. Not. Roy. astr. Soc.,* **192**, 427-439.

[24] Cosmological Constraints on the Scale of Grand Unification, 1980. *Nucl. Phys. B,* **181**, 461-487 (with G G Ross).

[25] Grand Unification Theories and the Dirac Hypothesis, 1980. *Nature*, **287**, 566.

[26] Cosmological Baryon Annihilation, 1980. *Mon. Not. Roy. astr. Soc.*, **192**, 19P-23P.

[27] Coagulation with Fragmentation, 1981. *J. Phys. A,* **14**, 729-733.

[28] Generic Singularity Studies Revisited, 1981. *Phys. Lett. A,* **82**, 441-6 (with F J Tipler).

[29] Chaos in the Einstein Equations, 1981. *Phys. Rev. Lett.*, **46**, 963-967.

[30] Baryosynthesis and the Origin of Galaxies, 1981. *Nature*, **291**, 469-472 (with M S Turner).

[31] Inflation in the Universe, 1981. *Nature*, **292**, 35-38 (with M S Turner).

[32] The Growth of Anisotropic Structures in a Friedmann Universe, 1981. *Astrophys. J.*, **250**, 432-449 (with J Silk).

[33] Cosmological Constraints on Elementary Particle Masses in Lemaître Cosmological Models, 1981. *Phys. Lett., B* **107**, 358-362.

[34] Cosmology and High Energy Physics, 1981. *Proc. Inst. of Physics Conference 'Nuclear Structure and Particle Physics'*, A Hilger, Bristol.

[35] The Lore of Large Numbers: Some Historical Background to the Anthropic Principle, 1981. *Quart. Jl. Roy. astr. Soc.,* **22**, 388-420.

[36] General Relativity and Non-linear Dynamics, 1982. *Gen. Rel. Gravn.*, **14**, 1523-1530.

[37] Chaotic Behaviour in General Relativity, 1982. *Phys. Reports* **85**, 1-49.

[38] Cosmology and Particle Physics, 1982. In 'Cosmology: an Oxford Symposium', ed. A Wolfendale, Reidel, Dordrecht, pp21-38.

[39] Dissipation and Unification, 1982. *Mon. Not. Roy. astr. Soc.,* **199**, 45P-48P.

[40] Massive Particles as a Probe of the Early Universe, 1982. *Nucl. Phys. B,* **208**, 501-508.

[41] Anthropic Definitions, 1983. *Quart. Jl. Roy. astr. Soc.*, **24**, 146-153.

[42] Natural Units before Planck, 1983. *Quart. Jl. Roy. astr. Soc.*, **24**, 24-26.

[43] Neutrino Proliferation, 1983. *Phys. Lett. B,* **125**, 377-378.

[44] Cosmology and Elementary Particles, 1983. *Fund. Cosmic Physics,* **8**, 83-200.

[45] Perturbations of a De Sitter Universe, 1983. In 'The Very Early Universe', ed. G Gibbons, S W Hawking and S T C Siklos, Cambridge University Press, Cambridge, pp267-272.

[46] Chaos in the Mixmaster Universe, 1983. *Phys. Rev. Lett.*, **50**, 134-137 (with D. Chernoff).

[47] The Stability of General Relativistic Cosmological Theory, 1983. *J. Phys. A,* **16**, 2757-2776 (with A. C. Ottewill).

[48] Limits on the Neutrino Number and Baryon Density of a Realistic Universe, 1983. *Mon. Not. Roy. astr. Soc.*, **203**, 393-401 (with J Morgan).

[49] Monopoles in the Milky Way, 1983. *Phys. Lett. B*, **122**, 31-32 (with J Stein Schabes).

[50] The Isotropy of the Universe, 1982. *Quart. Jl. Roy. astr. Soc.*, **23**, 344-357.

[51] Structure of the Cosmic Microwave Background, 1983. *Nature*, **305**, 397-402 (with R Juszkiewicz and D H Sonoda).

[52] Dimensionality, 1983. *Phil. Trans. Roy. Soc. A*, **310**, 337-346.

[53] Cosmology, Elementary Particles, and the Regularity of the Universe. In 'Relativistic Astrophysics and Cosmology', 1984. eds. X Fustero and E Verdaguer, World Publ. Singapore pp137-199.

[54] Non-linear Gravitational Clustering, 1984. *Mon. Not. Roy. astr. Soc.*, **209**, 19-144 (with R Juszkiewicz and D H Sonoda).

[55] Percolation Analyses of Observed and Simulated Galaxy Clustering, 1983. *Mon. Not. Roy. astr.Soc.,* **205**, 61P-66P (with S P Bhavsar).

[56] Inhomogeneous Cosmologies with Cosmological Constant, 1984. *Phys. Lett. A,* **103**, 315-318 (with J Stein-Schabes).

[57] Percolation Studies of Galaxy Clustering, 1984. In 'Clusters and Groups of Galaxies', eds. M Mardirossian, G Giuricin, and M Mezetti, Reidel, Dordrecht (with S P Bhavsar), pp 415-421.

[58] New Light on Heavy Light, 1984. *Nature*, **307**, 14-15 (with R Burman).

[59] Microwave Sky Patterns and the Density of the Universe, 1984. In *Proc. 1st ESO-CERN Symposium,* ed. G Setti and L van Hove, CERN, Geneva (with R Juszkiewicz and D H Sonoda) pp170-176.

[60] Chaotic Behaviour and the Einstein Equations, 1984. In 'Classical General Relativity', eds. W Bonnor, J Islam and M A H MacCallum, CUP, Cambridge pp 25-41.

[61] A Bootstrap Resampling Analysis of Galaxy Clustering, 1984. *Mon. Not. Roy. astr. Soc.*, **210**, 19P-23P, (with S P Bhavsar and D H Sonoda).

[62] The Invisible Universe, 1984. *New Scientist*, August 30, p 28-31 (with J Silk).

[63] From Quark to Cosmos, 1984. *Nature*, **308**, 13-14 (with J Silk).

[64] The Stability of Certain Spatially Homogeneous Cosmological Models, 1985. *Gen. Rel. Gravn.*, **17**, 409-416 (with D H Sonoda).

[65] Helium Formation in Cosmologies with Anisotropic Curvature, 1984. *Mon. Not. Roy. astr. Soc.*, **211**, 221-227.

[66] Cosmology, the Existence of Observers and Ensembles of Possible Universes, 1985. In 'Les Voies de la Connaissance', ed. M Cazenave, A Michel, Paris.

[67] First-Ranked Galaxies in Groups and Clusters, 1985. *Mon. Not. Roy. astr. Soc.,* **213**, 857-869 (with S P Bhavsar).

[68] Universal Rotation: How Large Can it be?, 1985. Mon. Not. Roy. astr. Soc., 213, 917-943 (with R Juszkiewicz and D H Sonoda).

[69] Closed Universes: Their Evolution and Final State, 1985. *Mon. Not. Roy. astr. Soc.,* **216**, 395-402 (with F J Tipler).

[70] Microwave Background Spottiness, 1985. *Nature*, **316**, 48 (with R Juszkiewicz and D H Sonoda).

[71] The Asymptotic Behaviour of Bianchi Universes, 1986. *Phys. Reports*, **139**, 1-49 (with D H Sonoda).

[72] The Limits of Observational Cosmology, 1985. *Observatory*, **105**, 120.

[73] Minimal Spanning Trees, Filaments and Galaxy Clustering, 1985. *Mon. Not. Roy. astr. Soc.,* **216**, 17-35 (with S P Bhavsar and D H Sonoda).

[74] Kaluza-Klein Mixmaster Universes, 1985. *Phys. Rev.* **D32**, 1595-8 (with J Stein-Schabes).

[75] The Stability of Kaluza-Klein Cosmologies, 1986. *Phys. Lett. B,* **167**, 173-177 (with J Stein-Schabes).

[76] General Relativistic Cosmological Models and the Cosmic Microwave Background, 1986. *Canadian J. Phys.*, **64**, 152-159.

[77] Alignments of Clustered Quasars, 1985. *Mon. Not. Roy. astr. Soc.,* **218**, 587-592 (with K Subramanian).

[78] The Left Hand of Creation: the origin and evolution of the expanding universe, (with J Silk), Basic Books, New York, pp 256.

[79] Three-Dimensional Classical Space-Times, 1985. *Classical & Quantum Gravity*, **3**, 551-567 (with A Burd and D Lancaster).

[80] The Inflationary Universe – Birth, Death, and Transfiguration, 1982. *Nature* **298**, 801-5 (with M S Turner)

[81] Observational Limits on the Time-Evolution of Extra Spatial Dimensions, 1987. *Phys. Rev. D,* **35**, 1805-1810.

[82] Filaments: What the Astronomer's Eye Tells the Astronomer's Brain, 1987. *Quart. Jl. Roy. Astr. Soc.*, **28**, 109-128 (with S P Bhavsar).

[83] Chaotic Space-times, 1987. In 'The Physics of Phase Space', Proc. 1st International Conference on Phase Space, pp 18-21, eds. A Zachary & S Kim, *Springer LNP Series* no. **278**, Springer, NY.

[84] The Closed-Universe Recollapse Conjecture, 1986. *Mon. Not. Roy. astr. Soc.,* **223**, 835-844 with G J Galloway & F J Tipler).

[85] Uncertainties in the Cluster-Cluster Correlation Function, 1986 . *Mon. Not. Roy. astr. Soc.,* **223**, P21-P27 (with E N Ling & C Frenk).

[86] Topics in Relativistic Cosmology, 1987. *Cargèse Lectures in Physics,* pp 239-292, eds. B Carter & J Hartle, Plenum, NY.

[87] Relativistic Cosmology, 1987. Vancouver Lectures. In 'The Early Universe', pp125-202, eds. W Unruh & G Semenoff, Plenum, NY.

[88] The Deflationary Universe: An Instability of the De Sitter Universe, 1987. *Phys. Lett. B,* **180**, 335-339.

[89] The Homogenisation and Isotropisation of an Inflationary Cosmological Model, 1986. *Phys. Lett. B,* **182**, 25-27 (with O Gron).

[90] Deflationary Universes with Quadratic Lagrangians, 1987. *Phys. Lett. B,* **183**, 285-289.

[91] Filaments and Cells – Reality of Illusions?. *Bull. Amer. Astronomical Soc.* Abstracts of 169th Meeting (with S P Bhavsar). 1987

[92] The Inflationary Universe: Modern Developments, 1988. *Quart. Jl. Roy. astr. Soc.,* **29**, 101-117.

[93] Inflation and No-hair Theorems, 1987. Proc. Paris-Meudon Colloquium – String Theory, *Quantum Cosmology and Quantum Gravity*, pp 42-53, eds. H de Vega & N Sanchez, World Scientific, Singapore.

[94] Cosmic No-hair Theorems and Inflation, 1987. *Phys. Lett. B*, **187**, 12-16.

[95] Non-gaussian Statistics and the Microwave Background Radiation, 1987. *Mon. Not. Roy. astr. Soc.*, **228**, 407-426 (with P Coles).

[96] Life, the Universe and the Anthropic Principle, 1987. *The World and I*, **2**, 179-187.

[97] The Premature Recollapse Problem in Closed Inflationary Universes, 1988. *Nucl. Phys. B*, **296**, 697-709.

[98] The Confrontation of No-scale Supergravity with Cosmology, 1988. *Nucl. Phys. B,* **298**, 733-747 (with N Deruelle).

[99] Chairman's Introduction to Royal Society Discussion meeting on the Material Content of the Universe, 1986. *Phil. Trans. Roy. Soc. A,* **320**, 433-34.

[100] Introduction to Relativistic Cosmology, 1987. In 'Proc. of the 1st CERN-ESO School on Astroparticle Physics, Erice, pp 487-571, ed. A De Rujula, World Scientific, Singapore.

[101] The Inflationary Universe, 1988. In 'Interactions and Structures in Nuclei', pp 135-150, eds. R Blin Stoyle & W D Hamilton, Adam Hilger, Bristol.

[102] Action in the Universe, 1988. *Nature*, **331**, 31-34 (with F J Tipler).

[103] Lensing of Supernova Neutrinos?, 1987. *Nature*, **327**, 375 (with K Subramanian).

[104] Inflationary Universes with Exponential Potentials, 1988. *Nucl. Phys. B,* **308**, 929-945 (with A Burd).

[105] String-driven Inflationary and Deflationary Universes, 1988. *Nucl. Phys. B*, **310**, 743-763.

[106] The Mysterious Lore of Large Numbers, 1988. Contributed chapter of 'Modern Cosmology in Retrospect', ed. B Bertotti, R Balbinot, S Bergia & A Messina, CUP, Cambridge, pp 67-92.

[107] Cosmology Limits the Number of Neutrinos, 1989. New Scientist 28th October, p 30 (with N Dombey)

[108] Inflation and the Conformal Structure of Higher-order Gravity Theories, 1988. *Phys. Lett. B,* **214**, 515-518 (with S Cotsakis).

[109] No New Kasner Solution in Higher-Derivative Gravity, 1989. *Mod. Phys. Lett. A*, **4**, 519.

[110] The New Cosmology, 1988. *Beshara* **6**, 19-25.

[111] Constants of Physics and the Structure of the Universe, 1988. In 'Unites de Mesure et Constantes Physiques', 30th Saas Fee Lectures, Chapter 8, pp 1-29, eds. M Batato et al., Assoc. Vaudoise, Lausanne.

[112] What is the Principal Evidence for the Cosmological Principle?, 1989. *Quart. J. Roy. astr. Soc.*, **30**, 163-167.

[113] Newtonian No-hair Theorems, 1989. *Class. Quantum Gravity*, **6**, 1253-1265 (with G Götz)

[114] *The Mathematical Universe*, 1989. World & I, May, pp306-311.

[115] Mixmaster Cosmological Model in Theories of Gravity with a Quadratic Lagrangian, 1989. *Phys. Rev. D*, **39**, 2187-92 (with H Sirousse-Zia).

[116] Microwave Background Constraints on the Voronoi Model of Large-scale Structure, 1990. *Mon. Not. Roy. astr. Soc.,* **244**, 557-562 (with P Coles).

[117] New Patterns of Explanation in Cosmology, 1989. Venice Conference on Philosophy and Cosmology, ed. U Curi, CUP, pp1-15.

[118] The Asymptotic Approach to De Sitter Space-time, 1989. *Phys. Lett. B,* **231**, 228-230 (with G Götz).

[119] Cosmological Theories: An Overview, 1992. *The Astronomy and Astrophysics Encyclopedia*, ed. S P Marran, pp 168-9, Van Nostrand, NY and Cambridge UP.

[120] Equilibrium Matter Fields in the Early Universe, 1990. *Mon. Not. Roy. astr. Soc.*, **243**, 602-609 (with W S Hawthorne).

[121] String-driven Inflation, 1990. In 'The Formation and Evolution of Cosmic Strings', eds. G Gibbons & T Vasasparti, pp 449-464, CUP, Cambridge.

[122] Inner Space and Outer Space: The Quest for Ultimate Explanation, 1989. *The Centenary Gifford Lectures*, ed. N Spurway, Chapter 3, pp 48-103, Blackwell, Oxford.

[123] The Inflationary Universe, 1989. In 'New Ideas in Cosmology' 1989, ed. M Shallis, Blackwell, Oxford.

[124] The Statistics of Primordial Density Perturbations, 1990. *Mon. Not. Roy. astr. Soc.*, **244**, 188-193 (with P Coles).

[125] Chaotic Behaviour in Higher-order Lagrangian Theories of Gravity, 1989. *Phys. Lett. B*, **232**, 172-176 (with S Cotsakis).

[126] Graduated Inflationary Universes, 1990. *Phys. Lett. B*, **235**, 40-43.

[127] De Sitter Ground States and Boundary Terms in Generalised Gravity, 1989. *Nucl. Phys. B*, **323**, 242-252 (with M Madsen).

[128] Platonic Relationships in the Universe, 1991. New Scientist, April 20th, pp40-43.

[129] Why is the World Mathematical? Scienza & Technica Annuario Astronomical, *Astrofisica e Cosmologica*, 1990, pp 10-17.

[130] Extended Inflationary Universes, 1990. *Nucl. Phys. B* **341**, 294-308 (with K Maeda).

[131] Projected and Intrinsic Shapes of Galaxy Clusters, 1990. *Mon. Not. Roy. astr. Soc.*, **249**, 662-677 (with M Plionis and C Frenk).

[132] Primordial Density Fluctuations and the Microwave Background Spectrum, 1991. *Mon. Not. Roy. astr. Soc.*, **248**, 52-57 (with P Coles).

[133] Baryogenesis in Extended Inflation I: Baryogenesis via Production and Decay of Supermassive Bosons, 1991. *Phys. Rev. D* **43**, 977-983 (with E Copeland, E W Kolb & A R Liddle).

[134] Baryogenesis in Extended Inflation II: Baryogenesis via Primordial Black Holes, 1991. *Phys. Rev. D* **43**, 984-994 (with E Copeland, E W Kolb & A R Liddle).

[135] Why is the Universe Mathematical?, 1990. In the Proceedings of 'Truth in Science', ed. V Erspamer, Acad. Licencei, Rome.

[136] Life, the Universe and the Anthropic Principle, In 'The Anthropic Principle', 1990. ed. J Taylor, Penguin.

[137] The Behaviour of Intermediate Inflationary Universes, 1990. *Phys. Lett B* **249**, 406-410 (with P Saich).

[138] Why is the Universe Mathematical, 1990. *Nuovo Civilta delle Macchine* **4**, 175-183.

[139] The Origin of the Universe, 1991. In 'La Narrazione delle Origini', ed. L Preta, Laterza, Rome.

[140] The Evolution of the Universe, 1991. *New Literary History* **22**, 835-856.

[141] The Evolution of Black Holes in an Expanding Universe, 1991. *Mon. Not. Roy. astr. Soc.*, **253**, 675-682 (with E Copeland & A R Liddle).

[142] Some Statistical Problems in Cosmology, 1992. In 'Statistical Problems in Astronomy', eds. A Feigelson and G J Babu, pp 21-45, Springer, NY.

[143] Self-regenerating Inflationary Universes in Higher-order Gravity in Arbitrary Dimension, 1991. *Phys. Lett. B* **258**, 299-304 (with S Cotsakis).

[144] Galaxy Clustering, 1992. In 'Statistical Problems in Astronomy', eds. A Feigelson and G J Babu, pp 163-170, Springer, NY.

[145] Making Machines of Gods, 1991. The Times Higher Ed. Supplt., June 1st issue, pp13-14.

[146] Baryosynthesis, Gravitation and Hot Big Bang Cosmology, 1992. In 'Latin American Winter School Lectures in Theoretical Physics', ed. D Altschuler, J Nieves, J Ponce de Leon & M Urbriaco, pp 171-214, World Scientific, Singapore.

[147] The Entropy of the Universe, 1991. In 'Evolutionary Phenomena in the Universe', ed. P Giannone, F Melchiorri and F Occhineiro, pp 203-210, Italian Physical Society, *Conf. Proc.*, **32**, Bologna.

[148] The Cosmology of Black Hole Relics, 1992. *Phys. Rev. D* **46**, 645-657 (with E Copeland & A R Liddle).

[149] Gravitation and Hot Big Bang Cosmology, 1991. In 'The Physical Universe: The Interface Between Cosmology, Astrophysics and Particle Physics', eds. JD Barrow, A Henriques, M Lago & M Longair, pp1-20, Springer-Verlag.

[150] Consequences of Extended Inflation, 1992. GIFT Lectures, Proc. St Feliu Lecture School, eds. J Sola et al., pp 179-190, World Scientific, Singapore.

[151] The Anthropic Principle, 1992. *L'Umana Avventura*, Feb. issue, pp1-13.

[152] Galaxy Clustering, 1992. In 'Statistical Problems in Astronomy', eds. A Feigelson and G J Babu, pp163-170, Springer-Verlag, NY.

[153] Theories of Everything: the Quest for Ultimate Explanation, 1991, pp223, Oxford UP, Oxford.

[154] Perche il Mondo e Matematico? 1991. In 'Scienza & Technica Yearbook, pp10-15, Mondadori, Milan.

[155] L'Origine dell'Universo, 1991. In 'La Narrazione delle Origini', ed. L Preta, pp 5-29, Laterza, Rome.

[156] Nuovi Modelli di Spiegazione in Cosmologi, 1991. In 'Il Principio Antropico', ed. B Giacomini, pp17-32, Spazio Libri Editori, Ferrera.

[157] Pi in the Sky: Counting, Thinking and Being, 1992. pp317, Oxford UP, Oxford and NY.

[158] L'Universo e Matematics?, 1992. *Sfera* **29**, 6-20.

[159] Stasera tutti do Euclide, 1992. La Repubblica Cultura, Dec. 20th, pp2-3.

[160] Shaking the Foundations of Mathematics, 1992. *New Scientist*, No. **1848**, pp 36-39.

[161] Gravitational Memory, 1992. *Physical Review D* **46**, R3227-3230.

[162] Tempo Cosmologico, 1995. In 'Il Temponella Scienza e nella Filosophia', ed. E Agazzi, pp 97-120, Guida Editori, Naples.

[163] Scalar-field Cosmologies, 1993. *Class. Quantum. Gravity* **10**, 279-283 (with P Saich).

[164] Gravitational Collapse of Rotating Pancakes, 1993. *Class. Quantum. Gravity* **10**, 79-91 (with P Saich).

[165] Die Entwicklung des Universums, 1993. In 'Neue Kosmologische Perspektiven', Insel Jahrbuch 1993, pp 44-71, ed. E Sens, Insel Verlag, Frankfurt.

[166] Life and the Big Bang, 1993. In 'Proc. Varenna School of Physics', ed. E Sindoni, pp1-19, Soc. Ital, di Fisica, Bologna.

[167] The Origin of the Universe, pp 150, Orion, London and Basic Books, NY.

[168] Grande Quanto Basta, 1993. *Sfera*, **40**, 8-35.

[169] Cosmological Principles, 1993. In 'The Renaissance of General Relativity and Cosmology', Proc. D W Sciama 65th Birthday Meeting, eds. G F R Ellis, A Lanza and J Miller, pp 201-212, Cambridge UP, Cambridge.

[170] The Growth of Large-scale Structure with a Cosmological Constant, 1993. *Mon. Not. Roy. astr. Soc.*, **262**, 717-725 (with P Saich).

[171] Scalar-tensor Cosmologies, 1992. *Physical Review D* **47**, 5329-5335.

[172] Perturbation Spectra From Intermediate Inflation, 1993. *Physical Review D* **47**, R5219-5223 (with A R Liddle).

[173] Relativistic Cosmology and the Regularization of Orbits, 1993. *Observatory* **113**, 210-211.

[174] Unprincipled Cosmology, 1993. *Quart. Jl. Roy. astr. Soc.*, **34**, 117-134.

[175] New Types of Inflationary Universe, 1993. *Phys. Rev. D* **48**, 1585-1590.

[176] In the World's Image, 1993. Times Higher Ed. Supplt., pp 15-16, 16th April

[177] Sizing Up the Universe, 1993. *Focus Oct.*, pp70-73.

[178] Black Hole Memory, 1993. *Gen. Rel & Gravn.*, **26**, 1-5.

[179] The Universe Began in no Time at all, 1993. *Falmer*, **22**, 12-13.

[180] Anthropic Principles in Cosmology, 1993. *Vistas in Astronomy*, **37**, 409-427.

[181] Energia e Materia, 1993. *Sfera*, **34**, 16-20.

[182] Non-singular Scalar-tensor Cosmologies, 1993. *Phys. Rev. D* **48**, 3592-3595.

[183] Amplification of Gravitational Waves in Scalar-tensor Theories of Gravity, 1993. *Phys. Rev. D* **48**, 3630-3640 (with J P Mimoso and M G Maia).

[184] Gravitational Memory, 1993. *Ann. NY Acad. Sci.*, **688**, 686-689.

[185] La Vita e il Big Bang, 1993. In 'Scienza ed Etica Alle Soglie del Terzo Millenio', *Proc. Soc. Ital. di Physica,* **36**, 141-156.

[186] Perfect Fluid Scalar-tensor Cosmologies, 1994. *Phys. Rev. D* **50**, 3746-3754 (with J P Mimoso).

[187] Exact Inflationary Universes with Potential Minima, 1994. *Phys. Rev. D* **49**, 3055-3058.

[188] Cosmological Graviton Production in general Relativity and Related Gravity Theories, 1994. *Phys. Rev. D* **50**, 6262- (with M G Maia).

[189] Amplification of Gravitational Waves in Scalar-tensor Theories of Gravity, 1993. *Phys. Rev. D* **3630**-3640 (with J P Mimosa and M G Maia).

[190] De Allesomvattende Verklaring, 1993. In 'Een goed begin is het hele werk: Beschouwingen over het begin van de kosmos', pp.95-106, Univ. Amsterdam, Amsterdam.

[191] An Explosion into Time and Space, 1994. Independent Supplt., 2nd Oct., 66.

[192] Caso è determinismo, 1994. In 'Il Caso è la Libertia, eds. M Ceruti et. al., pp 15-27,Laterza, Rome.

[193] Theories of Everything, 1994. In Physics and Our View of the World, ed. J. Hilgevoord, pp 38-61, Cambridge UP, Cambridge.

[194] Il Suono dei numeri, 1994. La Repubblica, 4th March, 36.

[195] Long-distance Calls, 1994. Guardian, 13th Oct., 13.

[196] Discussion, 1994. In 'Physics and Our View of the World', ed. J Hilgevoord, p 259-61, Cambridge UP, Cambridge.

[197] Vita, morte, è miracoli del i'universo, 1994. La Repubblica, 19th Sept., 28.

[198] Formalising the Slow-roll Approximation in Inflation, 1994. *Phys. Rev.* **D50**, 6262 (with A R Liddle and P Parsons).

[199] Why the Universe is not Anisotropic, 1995. *Phys. Rev.* **D51**, 3113-3116.

[200] Slow-roll Inflation in Scalar-tensor Theories, 1995. *Phys. Rev.* **D51**, 2729-2732.

[201] Oscillating Universes, 1995. *Mon. Not. R. astr. Soc.,* **275**, 850-862 (with M Dabrowski).

[202] Inflationary Models with Logarithmic Potentials, 1995. *Phys. Rev.* **D52**, 5576-5588 (with P Parsons).

[203] Generalised Scalar Field Potentials and Inflation, 1995. *Phys. Rev.* **D51**, 6757-6764 (with P Parsons).

[204] New Inflation from Old, 1995. *Class. Quantum Grav.*, **12**, 1715-1721 (with P Parsons).

[205] How Anisotropic can a Universe be?, 1995. *Ann. NY Acad. Sci.*, **59**, 706-709.

[206] Theories of Everything, 1995. In 'Nature's Imagination', ed.J Cornwell, pp 45-63, Oxford UP, London & New York.

[207] I Limiti della Conoscenza, 1995. In 'L'Uomo: I Limiti, Le Speranze', eds. G Giorello and E Sindoni, pp 105-111, Piemme, Casale Monferrato.

[208] Limiti della Scienza, 1995. In 'Terzo Millennio', Proc. Int. School Plasma Physics, Varenna, eds. G Giorello and E Sindoni, pp 101-112, New Press, Como.

[209] Stars in their Ancient Eyes, 1995. Times Higher Ed. Suppl., Dec 16th, p16.

[210] Musical Cheers and the Beautiful Noise, 1995. Times Higher Ed. Suppl., June 23rd, 19-20.

[211] Beauty is in the Eye of the Hunter, 1995. Times Higher Ed. Suppl., Sept. 15th, 18-19.

[212] The Wigner Inequalities for a Black Hole, 1996. *Phys. Rev. D* **54**, 6563-4.

[213] Varieties of Expanding Universe, 1996. *Classical & Quantum Grav.* **13**, 2965-2975

[214] Time-varying G, 1996. *Mon. Not. R. astr. Soc.*, **282**, 1397-1406

[215] Cosmological Models of Dimensional Segregation, 1996. *Classical & Quantum Grav.* **13**, 2693-2706 (with J. Yearsley).

[216] The Formation and Evaporation of Primordial Black Holes in Scalar-Tensor Gravity Theories, 1996. *Phys. Rev.* **D54**, 3920-3931 (with B.J. Carr),

[217] Limits of Science, 1996. In Boundaries and Barriers: on the limits to scientific knowledge, eds.J. Casti & A Karlqvist, pp.1-11, Addison Wesley, Reading Mass.

[218] The Behaviour of Cosmological Models with Varying G, 1997. *Phys. Rev. D* **54**, 1906-1936 (with P. Parsons)

[219] Spatially Homogeneous String Cosmologies, 1997. *Phys. Rev. D* **55**, 623-30 (with K.E. Kunze)

[220] Kantowski-Sachs String Cosmologies, 1997. *Phys. Rev. D* **55**, 630-8 (with M. Dabrowski)

[221] Frontiers and Limits of Science, 1997. In How Large is God? ed. J.M. Templeton, pp. 203-216, Templeton Foundation Press, Philadephia.

[222] Cosmological Limits on Slightly Skew Stresses, 1997. *Phys. Rev. D* **55**, 7451-60

[223] Is the World Simple or Complex?, 1997, In Scienza e Realtà, ed. E. Mariani, pp. 35-56, I.P.E., Naples.

[224] Time in the Universe, 1997. In The Idea of Progress, eds A. Burgen, P. McLaughlin & J. Mittelstrass, Pp. 155-174, Walter De Gruyter & Co., Berlin and New York.

[225] Some Limits of Science and Scientists, 1997. In Proc. 3rd International Symposium on Conceptual Tools for Understanding Nature, Trieste 1995, ed G. Costa, G. Calucci, and M. Giogi, World Scientific, Singapore, pp.1-10.

[226] The Origin of the Universe, Proceedings of Reflections on the Birth of the Universe, Varenna, 25/9/96, ed E Sindoni

[227] Come i fisici immaginano sia avvenuta la nascita dell'Universo?, 1997. In La Favola dell'Universo, eds G.V. Coyne, G. Giorello, and E. Sindoni, *Piemme*, pp. 144-146, Casale Monferrato.

[228] Science: Limits, Boundaries, and Frontiers. In Il Patto col Diavolo, ed. P. Donghi, pp. 15-36, Laterza, Rome, 1997.

[229] Wigner Inequalities for a Black Hole, 1996. *Phys. Rev. D* **54**, 6563-4.

[230] Constraints on a Primordial Magnetic Field, 1997. *Phys. Rev. Lett.* **78**, 3610-3 (with P.G. Ferreira, and J. Silk)

[231] Flat Spots: Topological signatures of open universes, 1997. *Phys. Rev. Lett.* **79**, 974-7 (with J. Levin, E. Bunn, and J. Silk)

[232] Inhomogeneous String Cosmologies, 1997. *Phys. Rev. D* **56**, 741-752 (with K.E. Kunze)

[233] Can Inflation be Falsified?, 1997. *Gen. Rel. Gravitation,* **29**, 1501-1508 (with A. Liddle).

[234] Geodesics in Open Universes, 1997. *Phys. Lett. A* **233**, 169-174 (with J. Levin)

[235] Chaos in the Einstein-Yang-Mills Equations, 1998. *Phys. Rev. Lett.* **80**, 656-9 (with J. Levin)

[236] A Gauge-invariant Analysis of Magnetic fields in General Relativistic Cosmologies, 1997. *Classical & Quantum Grav.* **14**, 2539-2562 (with C.G. Tsagas)

[237] Is the World Simple or Complex?, 1999. in The Values of Science: Oxford Amnesty Lectures 1997, ed W. Williams, Oxford Amnesty International Lectures, Westview Press, NY, Pp 80-94.

[238] Rational Vote is Doomed, 1997. Daily Telegraph, 30 April.

[239] The Analogy of Nature, 1999. In Mission to Abisko Abisko Symposium, eds. J. Casti & A. Karlqvist, Perseus Books, NY, pp. 1-6.

[240] New Limits on the Variation of Physical Constants, 1998. *Mon. Not. R. Astron. Soc.* **295**, 457-462 (with M. J. Drinkwater, J. K. Webb and V.V. Flambaum)

[241] Limits on Cosmological Magnetic Fields and Other Anisotropic Stresses, Proc. 4th Paris Cosmology Colloquium, 1997, Pp196-222, eds. H. De Vega, N. Sanchez, World Scientific, Singapore, 1998.

[242] Magnetohydrodynamics in the Early Universe and the Damping of Non-linear Alfv èn waves, 1998. *Phys. Rev.* **D58**, 083502/1-18 (with K. Subramanian),

[243] Spherical Curvature Inhomogeneities in String Cosmology, 1998. *Phys. Rev* **D 57,** 2255-2263, (with K. Kunze)

[244] How the Universe Got Its Spots, 1999. *Phys. Rev. D* **58**, 123006-1-14 (with J. Levin, E. Scannapieco, G. De Gasperis and J. Silk).

[245] Is there chaos in low-energy string cosmology, 1998. *Phys. Rev.* **D57**, 1-19 (with M. Dabrowski)

[246] New Limits on the Variation of Physical Constants, 1997. IAP Paris Conference Proceedings (with M. J. Drinkwater, J. K. Webb and V. Flambaum)

[247] Varying G and other Constants, 1998. Proc. Erice Summer School Current Topics in Astrofundamental Physics: Primordial Cosmology, 1997, ed. N. Sanchez & A. Zichichi, World Scientific, Singapore, pp 269-305.

[248] The Origin of the Universe, 1997, *Modern Astronomer* **1** (8), 52-3.

[249] Limits on Cosmological Magnetic Fields and Other Anisotropic Stresses, 1998. Proc. Erice Summer School Current Topics in Astrofundamental Physics: Primordial Cosmology, 1997, ed. N. Sanchez & A. Zichichi, World Scientific, Singapore, pp 241-267

[250] Cosmological Variations in G and α, 1998, in Particle Cosmology, Proc. 3rd RESCEU Symposium on Particle Cosmology, Tokyo, Nov 10-13, 1997, ed. K. Sato, T. Yanagida, and T. Shiromizu, Universal Academic Press, Tokyo, pp. 221-236.

[251] The Radiation-Matter Transition in Jordan-Brans-Dicke Theory, 1998. *Phys. Rev. D* **58**, 027302-1-4 (with A.R. Liddle and A. Mazumdar).

[252] Evidence for Time Variation of the Fine Structure Constant?, 1999. *Phys. Rev. Lett.* **82**, 884-7 (with J.K. Webb, V.V. Flambaum, C.W. Churchill, and M.J. Drinkwater).

[253] La complessità è un mucchio di zucchero, 1998. *Liberal* **1** (issue 2) p174-5, 12 March, 1998

[254] Gauge invariant magnetic perturbations in perfect-fluid cosmologies, 1998. *Class. Quantum Gravity* **15**, 1-22 (with C. Tsagas).

[255] Gödel Universes in String Theory, 1998. *Phys. Rev. D* **58**, 103502-1-5 (with M. Dabrowski)

[256] Microwave Signals from Tangled Magnetic Fields, 1998. *Phys. Rev. Lett.* **81**, 3575-9 (with K. Subramanian)

[257] Only Connect, Liberal (1998)

[258] Stringy Gödel Universes, Gravity Essay (with M. Dabrowski)

[259] Cosmologies with Varying Light-Speed, 1999. *Phys. Rev.* **D59**, 043515, 1-8.

[260] The Beginning of the Universe, 1998. In Aspetti del Tempo, IPS Naples, 8-10 Oct. 1997, ed E. Mariani, pp 67-82.

[261] String Cosmology and Chaos, 1998. Proc. International Seminar Current Topics in Mathematical Cosmology, 30 March-4 April, 1998, Potsdam, eds. M. Rainer & H-J. Schmidt, World Scientific, Singapore, (with M. Dabrowski). Pp. 53-62.

[262] El Origen del Universo, 1998. *Revista de Occidente* **207**, 78-95

[263] Cosmology, Design Arguments, and the Anthropic Principle, 1998. Proc. SSQ Conference, Berkeley, June 8th 1998, CTNS.

[264] Anisotropic Stresses in an Inhomogeneous Universe, 1998. *Phys. Rev. D* **59**, 043502, 1-8, (with R. Maartens).

[265] Genio e Rugolatezza, 1998. In Il Libro del Salone: i testi, ed. E. Gianotti, Salone del Libro, Prosa S.c.r.l., Turin. Pp. 173-179.

[266] Varying-α theories and solutions to the cosmological problems, 1998. *Phys. Lett. B* **443**, 104-110 (with J. Magueijo).

[267] String Cosmology, 1998. *Chaos, Solitons & Fractals* **10**, 1-9 (with K.E. Kunze).

[268] Art and Science: Les Liaisons Dangereuses. In Art and Complexity, eds. J. Casti & A Karlqvist, Proc. Abisko Worshop 1998, Elsevier Amsterdam, (2003), pp. 1-20.

[269] Solutions to the Quasi-flatness and Quasi-lambda Problems, 1999. *Phys. Lett. B* **447**, 246-250 (with J. Magueijo).

[270] Cosmology and the Origin of Life, 1999.In Origine della Vita Intelligente Nell'Universo: The Origin of Intelligent Life in the Universe, Proc. Varenna Conference, Varenna, Sept. 1998, ed. E. Sindoni et al, Edizioni New Press, Como. Pp. 7-18.

[271] Solving the Flatness and Quasi-Flatness Problems in Brans-Dicke Cosmologies with varying Light Speed, 1999. *Class. Q. Gravity,* **16**, 1435-54 (with J. Magueijo).

[272] From Here to Infinity, Templeton Symposium on Accelerating Creativity, Washington DC, 15 Feb. 1999, pp1-15., ed. C. Harper.

[273] Is Nothing Sacred?, 1999. *New Scientist Issue* **2196**, 21 July, pp 28-32.

[274] Entropic Principles, 1999. *New Astronomy* **4**, 333-338.

[275] Spatial Variations of Fundamental Constants, 2001. *MNRAS* **322**, 585-8 (with C. O'Toole)

[276] Variations on a Constant Theme, 2000. *Gen. Rel. Gravitation*, **32**, 1111-1124.

[277] Cosmology, Life, and the Anthropic Principle, 2001. *Annals NY Acad. Sciences* vol. 950, 139-153, *Proc. AAAS Symposium Cosmic Questions*, April 14-16, 1999, Washington DC, ed. J.B. Miller.

[278] Els limits del Mon des la perspectiva de la physica, 1999. Nexus 23, 6-21 Catalunya Centre for Culture, Barcelona ed G. Sebastian. Talk delivered May 3, 1999.

[279] Fractals and scars on a compact octagon, 2000. *Class. Quantum Gravity* **17**, L61-8 (with J. Levin)

[280] An Entropic Cosmological Principle, 1999. *Mod. Phys. Lett. A* **16**, 1067-71.

[281] Some Exact non-vacuum Bianchi VI0 and VII0 instantons, 2000. *Class. Quantum Grav.* **17**,1435-45 (With Y. Gaspar and P. Saffin).

[282] Life, the Universe, But Not Quite Everything, Dec. 1999. *Physics World* **12**, 31-5.

[283] Grande quanto basta, 1999. In Zero e infinito: I principi di ordine che intervengono in natura, *Cuen, Naples.* Pp. 15-40.

[284] Le frontiere della scienza. Un debattito sulla vita, la coscienza, l'universo e I limiti della scienza, 1999. In Limiti e frontiere della scienza, ed. P. Donghi, Editori Laterza, Rome. Pp. 47-99 (with G. Cairns-Smith, J. Casti, P.C.W. Davies, & R. Penrose).

[285] Can a changing α explain the supernova results?, 2000. *Ap. J. Lett.* **532**, L87 (with J. Magueijo)

[286] Chaos and Order in a Finite Universe, 1999. *Proc. Of 'The Chaotic Universe', ICRA, Rome,* (with J. Levin).

[287] L'Origine dell'Universo, 2000. *Astronomia* **207**, 20-29.

[288] Can the Universe escape Eternal Acceleration?, 2000. *Mon. Not. Roy. Astron. Soc.* **316**, L41-4 (with R. Bean & J. Magueijo)

[289] The Millennium Mathematics Project, 2000 *RSA Journal* **1** (4) pp 4-6.

[290] The Twin Paradox in Compact Spaces, 2001. *Phys. Rev. A* **63**, 044104-0441 (with J. Levin)

[291] Mathematical Jujitsu: some informal thoughts about Gödel and physics, 2000. *Complexity* **5** (5), 28-34

[292] Possible Evidence for a variable fine structure constant from QSO absorption lines – I. Motivations, analysis and results, 2001. Mon. Not. Roy. Astron. Soc, 327, 1208-22, (with M.T. Murphy, J.K. Webb, V.V. Flambaum, V.A. Dzuba, C.W. Churchill, J.X. Prochaska, & A.M. Wolfe).

[293] Making Maths Count, Wavelength, 2000. Issue 27 (Oct), pp. 8-10.

[294] The Isotropy of Compact Universes, 2001. *Class. Quantum Gravity,* **18**, 1753-66 (with H. Kodama)

[295] The Big, the Small, and the In-between, 2001Project Syndicate J. Rose

[296] How Chaos Co-exists with Order, 2002. In Design and Disorder: Perspectives from Science and Theology, ESSSAT Proceedings, Lyons 2001, eds. N. Gregersen & U. Görman, T & T Clark, London and Continuum, NY, pp.11-30.

[297] Further Evidence for Cosmological Evolution of the Fine Structure Constant, 2001. Phys. Rev. Lett. 87, 091301(with J. K.Webb, M. Murphy, V. Flambaum, V. Dzuba, C. Churchill, J. Prochaska, & A. Wolfe).

[298] The Far, Far Future, 2001. In Proceedings Pontifical Academy, ed., G.F.R. Ellis, Templeton Foundation Press, Radnor, pp.

[299] Time in the Universe – a Traveller's Guide, 2002. Il Tempo Tra Scienza e Filosofia, Proc. Varenna conference 'The Faces of Time', Sept 25-28th 2000, ed. G. Giorello, E. Sindoni & C. Sinigaglia, edizioni Unicopli, Milan, pp.25-32.

[300] Bianchi VIII Empty Futures, 2001. *Class Quantum Gravity* **18**, 1-14 (with Y. Gaspar)

[301] Cosmological Evolution of Black Holes in Brans-Dicke Gravity, 2001. *Class. Quantum Gravity* **18**, 4717-23 (with N. Sakai)

[302] All Universes Great and Small, 2001. *Int. J Mod. Phys D* **10** 785-790 (with H. Kodama)

[303] Is a good long jumper a good high jumper?, 2001. *American Journal of Physics* **69**, 105-6.

[304] Scaling Solutions and Reconstruction of Scalar Field Potentials, 2001. *Phys. Rev. D* **64**, 127301 (with C. Rubano)

[305] How Mother Nature makes a super Father Time tick, 2001. *THES* No.**1501**, 24th August, p. 18-19

[306] Why is the Universe Mathematical? 2001. In Frontiers of Life, vol I *The Origins of Life*, pp. 125-138, Academic Press, NY.

[307] A simple varying-alpha cosmology, 2002. *Phys. Rev. Lett.* **88**, 031302 (with H.B. Sandvik & J. Magueijo).

[308] A non-constructive proof, *Mathematical Digest* **124**, 9 (2001)

[309] Kaluza-Klein Anisotropy in the CMB, 2002. *Phys. Lett B.* **532**, 153-158 (with R. Maartens).

[310] Magnetic Brane-worlds, 2001. *Class. Quantum Gravity* **19**, 155-172 (with S. Hervik)

[311] The Behaviour of varying-alpha cosmologies, 2002. *Phys. Rev. D* **65**, 063504 (with H. Sandvik and J. Magueijo)

[312] Anthropic reasons for non-zero flatness and Δ, 2002. *Phys Rev. D* **65**, 123501, (with H. Sandvik and J. Magueijo)

[313] Variations of Alpha in Space and Time, 2002. *Phys. Rev. D* **66**, 043515 (with J. Magueijo and H. Sandvik)

[314] Evolution of Black Holes in Brans-Dicke Gravity, 2001. (with N. Sakai), *Proc. 10th Workshop on General Relativity and Cosmology,* Osaka, Sept. 2000 (gr-qc/0012067).

[315] Viaggiare nel Tempo?, 2001. In Figli di Crono, ed. C. Martini, Cortina, Milan, pp. 51-68.

[316] Is it *e* or is it *c* ? Experimental Tests of Varying Alpha, 2002. *Phys Lett. B* **549**, 284-9 (with J. Magueijo and H. Sandvik).

[317] A Cosmological Tale of Two Varying Constants, 2002. *Phys. Lett. B* **541**, 201-10 (with J. Magueijo and H. Sandvik).

[318] Domande senza risposta, 2002. In Matematica e cultura, ed. M. Emmer, Springer-Verlag, Milan, pp. 13-26.

[319] Small-scale microwave background anisotropies due to tangled primordial magnetic fields, 2002. *MNRAS* **335**, L57-61 (with K. Subramanian).

[320] Cosmology – A Matter of All and Nothing, 2002. *Astronomy and Geophysics,* **43** (4) 8-15.

[321] The Weyl Tensor in Spatially Homogeneous Cosmological Models, 2002. *Class. Quantum Gravity,* **19**, 5173-5193 (with S. Hervik).

[322] Qualitative Analysis of Universes with Varying Alpha, 2002. *Class. Quantum Gravity,* **19**, 6197-6212 (with D.F. Mota)

[323] Infinities: Vivere in eterno, 2002. *La Rivista del Libri*, No. **7/8,** p 11-13.

[324] De la science des limites et les limites de science, 2002. In La complexité, vertiges et promesses: Dix-huit histoires de sciences d'aujourd'hui, ed. Réda Benkirane, chap XV, pp.311-328, Editions Le Pommier, Paris (2002).

[325] Enigma variations, 2002. *New Scientist* 7th Sept. No. 2359, pp. 30-33.

[326] Das Leben im Universum: gross, klein und komplex, 2002. In S.E.T.I.: die Suche nach dem Ausserirdischen, ed T.D. Wabbel, Beustverlag, Munich, pp. 37-52.

[327] What is Time? 2002. In Big Questions in Science, ed. H. Swain, J. Cape, London, pp. 21-32.

[328] Constants and Variations: from alpha to omega, 2003. Astrophysics and Space Sci. 283, 645-660. Plenary Talk. Proc. of JENAM 2002 Workshop on Variation of Fundamental Constants, Porto, Sept. 2002 also published in C.J.A.P. Martins, The Cosmology of Extra Dimensions and Varying Fundamental Constants, Kluwer, Dordrecht, (2003), pp. 207-222.

[329] Changing Alpha: some matters of gravity, 2002. *Int. J. Mod. Phys. D* **11**, 1615-9 (with J. Magueijo and H. Sandvik).

[330] Unusual Features of Varying Speed of Light Cosmologies, 2003. *Phys. Lett. B* **564**, 1-7.

[331] Gauge-invariant Perturbations of Varying-alpha Cosmologies, 2003. *Class. Quantum Gravity* **20**, 2045-2062 (with D.F. Mota).

[332] More Ado About Nothing, 2004. In Perché esiste qualcosa invece di nulla?: Vuoto, Nulla, Zero, ed. S. Moriggi &E. Sindoni, Itaca Libri, Ra., pp.11-28.

[333] The Stability of the Einstein Static Universe, 2003. *Class. Quantum Gravity* **20**, L155-L164 (with G.F.R. Ellis, R. Maartens & C. Tsagas).

[334] The Future of Tilted Universes of Bianchi Type, 2003. *Class. Quantum Gravity,* **20**, 2841-2854 (with S. Hervik)

[335] Small-scale CMB Polarisation anisotropies due to Tangled Primordial Magnetic Fields, 2003. *Mon. Not. Roy. Astron. Soc.* **344**, L31-35 (with K. Subramanian and T.R. Seshadri)

[336] Indefinite Information Processing in Ever-expanding Universes, 2003. *Phys. Lett. B* **566**, 1-7 (with S. Hervik)

[337] A Test of a Test for Chaos, 2003. *Phys. Rev. E* submitted (with J. Levin)

[338] The Copernican Principle in Compact Spacetimes, 2003. *Mon. Not. Roy. Astron. Soc.* **346**, 615-8 (with J. Levin).

[339] Quasi-spherical Gravitational Collapse in Any Dimension, 2004. *Gen. Rel. Gravitation*, **36**, 231-243 (with U. Debnath and S. Chakraborty)

[340] Glitch, 2003. *New Scientist,* 7th June issue, pp. 44-5.

[341] Varying Alpha in a More Realistic Universe, 2004. *Phys. Lett. B,* **581**, 141-6 (with D.F. Mota).

[342] The Long View, 2003. *The Observer Real Time Supplement*, pp. 32-35, 29th June issue.

[343] Mapping the Dark Energy with varying Alpha, 2003. *Phys. Lett. B*, **578**, 235-240 (with D. Parkinson and B. Bassett)

[344] Wie aus nichts etwas wird, 2004. In Die Macht der Zeichen, pp.131-155 Jewish Museum Berlin, eds. D. Tyradellis and M.S. Friedlander, DuMont

[345] The Dynamics and Stability of the Gödel Universe, 2004. *Class. Quantum Gravity* **21**, 1773-1789 (with C. Tsagas)

[346] Upstaging the Infinite, 2003. *New Scientist* 27th September, p. 34-37.

[347] The Gödel Brane, 2004. *Phys. Rev. D* **69**, 064007 (with C. Tsagas)

[348] Local and Global Variations of the Fine Structure Constant, 2003. *Mon. Not. Roy. Astron. Soc.* **349**, 291-302 (with D.F. Mota)

[349] Quando il Cielo si Colora di Nero, 2003. La Repubblica, Diario, p.49., 1st Oct.

[350] Mathematical Explanation, 2004. In Explanations, ed. J. Cornwell, Oxford UP, Oxford, pp. 81-109.

[351] Science in Culture: Gardening by numbers, 2004. *Nature* **427**, 296.

[352] Our Universe and Others, 2004. In Human Approaches to the Universe, *Proc. Academy of Finland Conference,* Helsinki, September 2003, ed. L. Larjo.

[353] Varying Alpha in Cosmology, 2004. *Mod. Phys Lett. A* **19**, 951-8, Proc. CosPA 2003, Taiwan.

[354] Marte: un pianeta tra it mito e la scienza, 2004. La Repubblica, Diario, p45, 7th Feb

[355] Sudden Future Singularities, 2004. *Class. Quantum Gravity* **21**, L79-L82.

[356] Chemistry and Sensitivity, 2004. In Fine Tuning in Chemistry, eds. J.D. Barrow, S. Conway-Morris, S. Freeland and C. Harper.

[357] Alles oder Nichts?: Die Vorstellungen vom Urknall und die Konstanten der Natur, 2004. In Mensch und Kosmos: Unser Bild des Universums, pp. 208-233, eds. E.P. Fischer and K. Wiegandt, Fischer Taschenbuch Verlag, Frankfurt, (2004).

[358] Outer Space, 2004. In Space In Science, Art and Society, ed. F. Penz, G. Radick & R. Howell, Cambridge UP, Cambridge, pp. 172-200.

[359] Eddington; leading the field, 2004. *Astronomy and Geophysics* **45**, 3.10-3.13 (with L. Mestel).

[360] Bouncing Universes with varying Alpha and varying G, 2004. *Class. Quantum Grav.* **21**, 4289-4296 (with D. Kimberly and J. Magueijo).

[361] Inhomogeneous Gravity, 2005. *Mon. Not. R. Astron. Soc.* **358**, 601-13 (with T. Clifton and D.F. Mota)

[362] Do Fermions and Bosons Produce the Same Gravitational Field?, 2004. *Phys. Rev. D* **70**, 103515 (with R.J. Scherrer)

[363] Outward Bound, 2005. *Physics Teacher* (Einstein Year Special Issue), Jan/March.

[364] Inconstant Constants, 2005. *Scientific American* (with J.K. Webb) June issue, pp. 56-63.

[365] At Home in a Goldilocks Universe, 2005. *Science and Spirit* , Jan/Feb issue, pp. 42-49.

[366] More General Sudden Singularities, 2004. *Class. Quantum Grav.* **21**, 5619-5622.

[367] Living in a Simulated Universe, 2005. In Universe or Multiverse, ed B.J. Carr, Cambridge UP.

[368] New Isotropic and Anisotropic Sudden Singularities, 2005. *Class. Quantum Grav.* **22**, 1563-1571 (with C.G. Tsagas)

[369] Structure and stability of Lukash plane-wave spacetimes, 2005. *Class. Quantum Grav.* **22**, 825-839 (with C.G. Tsagas).

[370] Doing An Infinite Number of Things Before Breakfast, 2005. *New Scientist* 29 Jan issue, pp. 28-32.

[371] Varying Couplings in Electroweak Theory, 2005. *Phys. Rev. D* **71**, 063525 (with D. J. Shaw).

[372] Einstein as Icon, 2005. *Nature* **433**, 218-291.

[373] The beginning of the universe (L'inizio dell'Universo), 2005. In Contesti e Validit`a delDiscorso Scientifico, eds. L. Cuccurullo & E. Mariani, Armando Editore, Rome, pp.38-65.

[374] Is the world simple or complex? (I mondo è semplice o complesso?), In Contesti e Validità delDiscorso Scientifico, eds. L. Cuccurullo & E. Mariani, Armando Editore, Rome, pp. 164-196.

[375] Varying Constants, 2005. *Phil Trans. Roy. Soc.* in press

[376] Cosmological Constants and Variations, 2005. *Journal of Physics*: Conference Series in press

[377] Cosmological Constraints on a Dynamical Electron Mass, 2005. *Phys. Rev. D* in press astro-ph/0503222 (with J. Magueijo)

[378] General Cosmological Bounds on Spatial Variations of Physical Constants, 2005. *Phys. Rev. D* **71**, 083520.

[379] Worlds without end or beginning, 2005. In The Scientific Legacy of Fred Hoyle, ed. D. Gough, Cambridge UP, pp.93-101.

[380] Benvenuti all'Albergo Infinito, 2005. In Lo Specchio di Carta (vol B) L'Espressione Lirica e Drammatica: Poesia e Teatro, ed. N. Botta, Loescher Editore, Turin, pp. 358-361.

[381] The Laws of Nature and the Flaws of Nature, 2005. Proc. Varenna School ed. E. Sindoni in press

[382] The Variation of G and Primordial Nucleosynthesis, 2005. *Phys. Rev. D* **71**, 123526 (with T. Clifton and R. Scherrer).

[383] Il nostro universo.. e tutti gli altri, 2005. KOS: rivista di medicina, cultura e scienze umane (publ. Istituto Scientifico Universitario San Raffaele, Milan), pp 22-27

[384] Bookends: Infinity and Beyond, 2005. Science and Theology July/August issue, pp.62-63.

[385] The Laws and Flaws of Nature, 2005. In Dio, la Natura e la Legge: God and the Laws of Nature, eds S. Moriggi & E. Sindoni, Angelicum-Mondo X, Milan., 88-92.

[386] Infinity – Dead or Alive, 2005. THES

[387] Outward Bound, 2005. In Spiritual Information, chap. 21, ed. C.L. Harper, Templeton Fdn Press Conshohocken, pp. 117-120.

Part RII - Books by John D. Barrow

B1 The Left Hand of Creation: The Origin and Evolution of the Expanding Universe, 1983. Pp. 256. Basic Books, New York and Heinemann, London (with J Silk); paperback eds by Allen & Unwin, London and Basic Bks., NY; 2nd enlarged edition 1995, pp 262, publ. Oxford UP, NY and Penguin, London + various languages.

B2 L'Homme et le Cosmos, 1984. Pp. 113, Radio France Imago, Paris (with F J Tipler).

B3 The Anthropic Cosmological Principle, 1986. Pp. 706, Oxford University Press, Oxford and New York+ various languages (with F J Tipler), 2nd edn. 1996.

B4 The World Within the World, 1988. Pp. 398, Oxford University Press, London and NY, hard and paper; 2nd edn 1995 and various languages.

B5 Theories of Everything: The Quest for Ultimate Explanation, 1991. Pp. 223, Oxford University Press, London and NY, and various languages. Paperback edns by Vintage, London and Ballantine, NY.

B6 Pi in the Sky: Counting, Thinking and Being, 1992. Pp. 317, Oxford University Press, London and NY, and various languages. Paperback edns by Penguin, London and Little Brown, NY.

B7 Perche il Mondo e Matematico?, 1992. Pp. 108, (1991 Sigma-Tau-Laterza Lectures) Laterza, Rome and various languages.

B8 The Origin of the Universe, 1993. Pp. 150, Basic Books, NY and Orion, London (and 26 languages).

B9 The Artful Universe, 1995. Pp. 274, Oxford University Press, London and NY and various languages. Paperback edns by Penguin, London and Little Brown, NY.

B10 Impossibility: the limits of science and the science of limits, 1998. Oxford U.P., Oxford and NY. Pp. 279 and various languages.

B11 Between Inner Space and Outer Space, 1999. Oxford U.P., Pp. 274 + various languages.

B12 The Universe that Discovered Itself, 2000. Oxford U.P., London and New York, pp. 448.

B13 The Book of Nothing, 2000. Jonathan Cape London & Pantheon NY (2001) + various languages, pp.361.

B14 Infinities, 2002. Editizione Piccolo Teatro di Milano, Milan in Italian, pp. 94.

B15 The Constants of Nature: From Alpha to Omega, 2002. Jonathan Cape, London and 2003 Pantheon NY+ various languages, pp. 353.

B16 The Infinite Book: A Short Guide to the Boundless, Timeless and Endless, 2005. Jonathan Cape London and Pantheon NY+ various languages, pp.328.

B17 The Artful Universe Expanded, 2005. 2nd edn Oxford UP, London and NY, pp. 321.

Part RIII - Other Authors' References

Abbot, L.F. (1985) - *Physics Letters B* **249**, 332.

Abdel-Rahman,A.-M.M. (1992) - *Physical Review D* **45**, 3497.

Abramo,L.R.W.; Brandenberger, R.H.;Mukhanov, V.F. (1997) - *Physical Review D* **56**, 3248.

Adler, R.; Bazin, M,; Schiffer, M. (1975) - *Introduction to General Relativity* - 2nd. edtn., McGraw-Hill, N.Y.

Adler, R.J.; Silbergleit, A.S. (2000) - *International Journal of Theoretical Physics*, **39**, 1291.

Aguirregabiria, J.M.; Chamorro, Virbhadra, K.S.; (1996) - *Gen. Rel. and Grav*. 28, 1393.

Albrecht,A.; Magueijo, J.(1998) *A time varying speed of light as a solution to cosmological puzzles*-preprint.

Arbab, A.I.(1997) - *GRG* **29**, 61.

Arbab, A.I.; Abdel-Rahman, A.-M.M. (1994) - *Physical Review D* **50**, 7725.

Bahcall, J.N.; Schmidt, M. (1967) *Physical Review Letters*, 19, 1294.

Banks, T. (1984) - *Physical Review Letters* **52**, 1461.

Banks, T. (1988) - *Nuclear Physics B* **309**, 493.

Barbour, J.; Pfister, H.; eds. (1995) - *Mach's Principle: From Newton's Bucket to Quantum Gravity*, Birkhuser, Boston.

Barr, S.M.(1987) - *Physical Review D* **36**, 1691.

Barrow, J.D. (1983). in *The Very Early Universe.*, Gibbons, G.N., Hawking, S.W., and Siklos, S.T.C., eds., Cambridge University Press, Cambridge.

Barrow, J.D. (1987) Phys. Rev., **D35**, 1805.

Barrow, J.D. (1990) - in *Modern Cosmology in Retrospect*. Edited by B.Bertotti, R. Balbinot, S.Bergia and A.Messina, Cambridge U.P., Cambridge.

Barrow, J.D. (1990 a) *Phys Lett.***B235**,40.

Barrow, J.D. (1991) *Theories of Everything*, Oxford UP, Oxford.

Barrow, J.D. (1992) *Phys.Rev* **D46**, R3227.

Barrow, J.D. (1993) *Phys. Rev* **D47**, 5329.

Barrow, J.D. (1993 a) *Phys.Rev.* **D48**, 3592.

Barrow, J.D. (1994) *The Origin of the Universe, Basic Books*, NY.

Barrow, J.D. (1995) Phys. Rev **D51**, 2729.

Barrow, J.D. (1996) MNRAS **282**, 1397.

Barrow, J.D. (1997) *Varying G and Other Constants*, Los Alamos Archives http://arxiv.org/abs/gr-qc/9711084 v1 27/nov/1997.

Barrow, J.D. (1998) Cosmologies with Varying Light Speed, Los Alamos Archives http://arxiv.org/abs/Astro–Ph 9811022 v1.

Barrow, J.D. (2002) - Los Alamos Archives http://arxiv.org/abs/gr-qc/0211074

Barrow, J.D. (2002) - Los Alamos Archives http://arxiv.org/abs/gr-qc/0209080

Barrow, J.D. (2004) - Los Alamos Archives http://arxiv.org/abs/gr-qc/0409062

Barrow, J.D. (2004) - Los Alamos Archives http://arxiv.org/abs/gr-qc/0403084

Barrow, J.D. (2005) - Los Alamos Archives http://arxiv.org/abs/astro-ph/0503434

Barrow, J.D. ; Papers in http://arXiv.org not cited above

Barrow, J.D. ; et al (2003) - Los Alamos Archives http://arxiv.org/abs/astro-ph/0307227

Barrow, J.D. ; et al (2003) - Los Alamos Archives http://arxiv.org/abs/gr-qc/0305075

Barrow, J.D. ; et al (2003) - Los Alamos Archives http://arxiv.org/abs/astro-ph/0303014

Barrow, J.D. ; et al (2003) - Los Alamos Archives http://arxiv.org/abs/gr-qc/0302094

Barrow, J.D. ; et al (2004) - Los Alamos Archives http://arxiv.org/abs/astro-ph/0406369

Barrow, J.D.;Carr, B.J. (1996) Phys. Rev **D54**, 3920.

Barrow, J.D.;Cotsakis S. (1988) Physics Letters **B214** 515.

Barrow, J.D.;Daborwski, M.P.(1995) – MNRAS **275**, 850.

Barrow, J.D.;Gotz, G. (1989) Class.Quantum Gravity **6**, 1253.

Barrow, J.D. ; Hervik, Sigbjorn (2002) - Los Alamos Archives http://arxiv.org/abs/gr-qc/0206061

Barrow, J.D. ; Hervik, Sigbjorn (2002) - Los Alamos Archives http://arxiv.org/abs/gr-qc/0302076

Barrow, J.D. ; Hervik, Sigbjorn (2003) - Los Alamos Archives http://arxiv.org/abs/gr-qc/0304050

Barrow, J.D. ; Levin, Janna (2003) - Los Alamos Archives http://arxiv.org/abs/gr-qc/0304038

Barrow, J.D.;Liddle, A.R.(1993) Phys. Rev **D47**, R5219.

Barrow, J.D.;Maeda, K.(1990) Nucl. Phys. **B341**, 294.

Barrow, J.D.;Magueijo J.(1999)– *Solving the Flatness and Quasi Flatness Problems in Brans-Dicke Cosmologies with a Varying Light Speed*, Los Alamos Archives http://arxiv.org/abs/astro-ph/9901049 5/01/1999.

Barrow, J.D. ; Magueijo, J. (2005) - Los Alamos Archives http://arxiv.org/abs/astro-ph/0503222

Barrow, J.D.;Mimoso, J.P.(1994) Phys. Rev **D50**, 3746.

Barrow, J.D. ; Mota, D. F. (2002) - Los Alamos Archives http://arxiv.org/abs/gr-qc/0207012

Barrow, J.D. ; Mota, D. F. (2002) - Los Alamos Archives http://arxiv.org/abs/gr-qc/0212032

Barrow, J.D.;Parsons, P. (1996) Phys. 6 Rev **D000**, 000.

Barrow, J.D.;Saich, P.(1990) Phys. Lett **B249**, 406.

Barrow, J.D. ; Scherrer, Robert, J.(2004) - Los Alamos Archives http://arxiv.org/abs/astro-ph/0406088

Barrow, J.D. ; Shaw, Douglas, J. (2004) - Los Alamos Archives http://arxiv.org/abs/gr-qc/0412135

Barrow, J.D. ; Subramanian, Kandaswamy (2002) - Los Alamos Archives http://arxiv.org/abs/astro-ph/0205312

Barrow, J.D. ; Tsagas, Christos, G.(2003) - Los Alamos Archives http://arxiv.org/abs/gr-qc/0309030

Barrow, J.D. ; Tsagas, Christos, G.(2003) - Los Alamos Archives http://arxiv.org/abs/gr-qc/0308067

Barrow, J.D. ; Tsagas, Christos, G.(2004) - Los Alamos Archives http://arxiv.org/abs/gr-qc/0411070

Barrow, J.D. ; Tsagas, Christos, G.(2004) - Los Alamos Archives http://arxiv.org/abs/gr-qc/0411045

Barrow, J.D.; Tipler, F.J. (1996) *The Anthropic Cosmological Principle*, Oxford University Press, Oxford.

Barrow, J.D.;Mimoso, J.P; Maria, M.R.G.(1993) Phys. Rev **D48**, 3630.

Beesham, A. (1986) - *International Journal of Theoretical Physics* **25**, 1295.

Beesham, A. (1993) - *Physical Review D* **48**, 3539.

Beesham, A. (1995) - GRG **27**, 15.

Bekenstein, J.D. (1974) - *Ann. Phys. (New York)* **82**, 535.

Bergmann, P.G. (1968) Int.J.Theor.Phys. **1**, 25.

Berman, M.S. (1983) - *Nuovo Cimento* **74B**, 182.

Berman, M.S. (1988) - GRG **20**, 841.

Berman, M.S. (1988 b) - GRG **21**, 967.

Berman, M.S. (1989) - GRG **21**, 967.

Berman, M.S. (1989) - *Physics Letters A* **142**, 335.

Berman, M.S. (1989 b) - Physics Letters **A 139**, 119.

Berman, M.S. (1989 d) - Physics Letters **A 142**, 227.

Berman, M.S. (1990) - *International Journal of Theoretical Physics* **29**, 1419.

Berman, M.S. (1990 a) - *International Journal of Theoretical Physics,* **29**, 567-570.

Berman, M.S. (1990 b) - GRG **22,**389.

Berman, M.S. (1990 c) - International Journal of Theoretical Physics **29**, 1415.

Berman, M.S. (1990 e) - International Journal of Theoretical Physics **29**, 571.

Berman, M.S. (1990 f) - Nuovo Cimento **B 105**, 1373.

Berman, M.S. (1990 g) - Nuovo Cimento **B 105**, 239.

Berman, M.S. (1990 h) - Nuovo Cimento **B 105**, 235.

Berman, M.S. (1990 i) - International Journal of Theoretical Physics **29**, 1423.

Berman, M.S. (1991) - *GRG* **23**, 465.

Berman, M.S. (1991 b) - *Physical Review D* **43**, 1075.

Berman, M.S. (1991 c) - *GRG* **23**, 1083.

Berman, M.S. (1992) - *International Journal of Theoretical Physics* **31**, 1447.

Berman, M.S. (1992 b) - *International Journal of Theoretical Physics* **31**, 321.

Berman, M.S. (1992 c) - *International Journal of Theoretical Physics* **31**, 1217.

Berman, M.S. (1992 d) - *International Journal of Theoretical Physics* **31**, 329.

Berman, M.S. (1992 e) - *International Journal of Theoretical Physics* **31**, 1451.

Berman, M. S. (1994) - *Astrophys. Space Science,* **222**, 235.

Berman, M.S. (1994 a) - *International Journal of Theoretical Physics* **33**, 1929.

Berman, M.S. (1994 c) - *Astrophysics and Space Science* **215**, 135. (by mistake, this journal published again Berman, 1992 c).

Berman, M.S. (1994 d) - *International Journal of Theoretical Physics* **33**, 1961.

Berman, M.S. (1996) - International Journal of Theoretical Physics **35**, 1789.

Berman,M.S. (1996 a) - International Journal of Theoretical Physics **35**, 1033.

Berman, M.S. (1996 b) - International Journal of Theoretical Physics **35**, 1719.

Berman, M.S. (1997) - International Journal of Theoretical Physics **36**, 1249.

Berman, M.S. (1997 a) - International Journal of Theoretical Physics **36**, 1461.

Berman, M.S. (2004) - Los Alamos Archives http://arxiv.org/abs/gr-qc/0407026

Berman, M.S. (2004 a) - Los Alamos Archives http://arxiv.org/abs/gr-qc/0412054

Berman, M.S. (2005) - Los Alamos Archives http://arxiv.org/abs/physics/0507110

Berman, M.S. (2006) - *Energy of Black-Holes and Hawking's Universe.* In *Trends in Black-Hole Research,* Chapter 5. Edited by Paul Kreitler, Nova Science, New York.

Berman, M.S. (2006 b) - *Energy, Brief History of Black-Holes, and Hawking's Universe.* In *New Developments in Black-Hole Research*, Chapter 5. Edited by Paul Kreitler, Nova Science, New York.

Berman, M.S (2006 c) - *On the Machian Properties of the Universe*, submitted to publication. Los Alamos archives: http://www.arxiv.org/abs/physics/0610003

Berman, M.S. (2006 d) - *On the Magnetic Field and Entropy Increase of a Machian Universe* - submitted to publication. Los Alamos Archives http://arxiv.org/abs/physics/0611007.

Berman, M.S. (2006 e) - *The Pioneer anomaly and a Machian Universe* - submitted to publication. Los Alamos Archives http://arxiv.org/abs/physics/0606117.

Berman, M.S. (2006 f) - Los Alamos Archives http://arxiv.org/abs/gr-qc/0605063

Berman, M.S. (2006 g) - Los Alamos Archives http://arxiv.org/abs/gr-qc/0605092

Berman, M.S. (2006 h) - Los Alamos Archives http://arxiv.org/abs/physics/0612007 to be published by *Astrophysics and Space Science.*

Berman, M.S. (2006 i) - Los Alamos Archives http://arxiv.org/abs/physics/0607005 to be published by *Rev. Mex. Astron. Astrofis.* 43, 2(2007)

Berman, M.S. (2006 j) - Los Alamos Archives http://arxiv.org/abs/physics/0608053

Berman, M.S. (2006 k) - Los Alamos Archives http://arxiv.org/abs/physics/0606208

Berman, M.S. (2006 l) - Los Alamos Archives http://arxiv.org/abs/physics/0609026

Berman, M.S. (2007) - *Introduction to General Relativistic and Scalar Tensor Cosmologies.*Nova Science, New York.

Berman, M.S. (2007 b) - *A Primer in Black Holes, Mach's Principle, and Gravitational Energy.*In preparation.

Berman, M.S.; Gomide, F.M. (1987) - *Tensor Calculus and General Relativity: an introduction.* (in Portuguese), McGraw-Hill, São Paulo, 2nd ed (first edition in 1986).

Berman, M.S; Gomide, F.M. (1988) - *GRG* **20**, 191.

Berman, M.S; Gomide, F.M. (2006) - *On a time variation of Neutrinos' mass* - submitted. See also Los Alamos archives: http://www.arxiv.org/abs/physics/0606208.

Berman, M.S.; Marinho Jr., R. M. (1996) - Letter to the Editor, *Physics Today,* **49,** 13.

Berman, M.S.; Marinho Jr., R. M. (1996 b) - Nuovo Cimento B, **111**, 1279.

Berman, M.S.; Marinho Jr., R. M. (2001) - *Astroph. Space Science* **278**, 367.

Berman, M.S.; Paim, T. (1990) - Nuovo Cimento B, **105**, 1377.

Berman, M.S.; Som, M.M. (1989) - *Physics Letters A* **142**, 338.

Berman, M.S; Som, M.M. (1989 b)Progress Theor.Phys, **81**, 823.

Berman, M.S; Som, M.M. (1989 c)Phys. Letters **A 136**, 206.

Berman, M.S.; Som,M.M. (1989 d). Nuovo Cimento **103B,N.2**, 203.

Berman, M.S.; Som, M.M. (1989 e) - Physics Letters **A 136**, 428.

Berman, M.S.; Som, M.M. (1989 f) - Physics Letters **A 139**, 119.

Berman, M.S.; Som, M.M. (1989 g) - GRG, **21**, 967-970.

Berman, M.S; Som, M.M. (1990) - *GRG* **22**, 625.

Berman, M.S.; Som, M.M. (1990 b) - *International Journal of Theoretical Physics* **29**, 1411.

Berman, M.S; Som, M.M. (1992) - International Journal of Theoretical Physics **31**, 325.

Berman, M.S.; Som, M.M. (1993) - Astrophysics and Space Science, **207**, 105.

Berman, M.S.; Som, M.M. (1993 b) - Journal of Mathematical Physics, **34(1)**, 111.

Berman, M.S; Som, M.M (1995) - Astrophys. Space.Sci, **225**, 237.

Berman, M.S; Som, M.M.; Gomide, F.M. (1989) - GRG **21**, 287.

Berman, M.S.; Trevisan, L.A. (2001) - Los Alamos Archives http://arxiv.org/abs/gr-qc/0104060

Berman, M.S.; Trevisan, L.A. (2001) - Los Alamos Archives http://arxiv.org/abs/gr-qc/0112011

Berman,M.S.; Trevisan, L.A. (2001 b) - Los Alamos Archives http://arxiv.org/abs/gr-qc/0111102

Berman, M.S.; Trevisan, L.A. (2001 c) - Los Alamos Archives http://arxiv.org/abs/gr-qc/0111101

Berman, M.S.; Trevisan, L.A. (2001 d) - Los Alamos Archives http://arxiv.org/abs/gr-qc/0111098

Berman, M.S.; Trevisan, L.A. (2001 e) - Los Alamos Archives http://arxiv.org/abs/gr-qc/0111099

Berman, M.S.; Trevisan, L.A. (2001 f) - Los Alamos Archives http://arxiv.org/abs/gr-qc/0111100

Berman, M.S.; Trevisan, L.A. (2002) - Los Alamos Archives http://arxiv.org/abs/gr-qc/0207051

Bernardis, P. et al. (2000), *Nature* **404**, 955;

Bertolami, O.(1986) - *Nuovo Cimento B* **93**, 36.

Bertolami, O. (1986 b). Fortschritte der Physik, **34**, (12), 829.

Bondi, H.; Gold, T. (1948). MNRAS, **108**, 252.

Bonnor, W.; Cooperstock, F.I. (1989) - *Phys. Lett. A* **139**, 442.

Boyer, R.H.; Lindquist, R.W. (1967) - Journal of Mathematical Physics, **8**, 265.

Brans, C.; Dicke, R. H. (1961) - *Phys. Review* **124**,925.

Brown, J.D.; Teitelboim, C.(1987) - *Physics Letters B* **195**, 177.

Burd A.B., Barrow J.D. (1988) Nucl. Physics, **B308**, 929.

Carmeli, M. (1982) - *Classical Fields* - Wiley, N.Y.

Carroll, S. M. (2001), *Living Rev. Rel.* **4**, 1.

Carvalho, J.C.; Lima, J.A.S.; Waga, I. (1992) - *Physical Review D* **46**, 2404.

Chaliassos, E. (1987). Physica, **144A**, 390.

Chen, W.; Wu, Y.- S. (1990) - *Phys. Review* **D 41**,695.

Ciufolini, I.; Pavlis, E. (2004) - *Letters to Nature*, **431,** 958.

Ciufolini, I. (2005) - Los Alamos Archives http://arxiv.org/abs/gr-qc/0412001 v3

Ciufolini, I.; Wheeler, J. A. (1995) - *Gravitation and Inertia*, Princeton Univ. Press, Princeton. See especially page 82, # 72 to 83.

Coleman, S. (1988) - *Nuclear Physics B* **310**, 643.

Cooperstock,F.I.;Israelit,M. (1995) - *Foundations of Physics,* **25**, 631.

Cooperstock, F.I.; Rosen, N. (1989) - *International Journal of Theoretical Physics,* **28**, 423 - 440.

Cooperstock, F.I.; Tieu, S. (2005) - submitted to Astrophysical Journal. See also Los Alamos Archives http://arxiv.org/abs/astro-ph/0507619.

Cusa, N. (1954) - *Of Learned Ignorance*, Routledge and Kegan Paul, London.

Dicke, R.H. (1959) - *Science* **129**, 621.

Dicke, R.H. (1962). *Physical Review*, **125**, 2163.

Dicke, R.H. (1964) - in *Gravitation and Relativity,* W.A.Benjamin Inc. New York, p1.

Dicke, R.H. (1964) - ibidem p.121.

Dicke, R.H. (1967) - *Physics Today* **20**, 55.

Dicke, R.H. et al. (1965) - *Ap. J.* **142**, 414.

Dickey, G. O.; Bendel P.L.; Faller, J.E.; Newhall, X.X.; Ricklefs, R.L.; Ries, J.G.; Shelus, P.J.; Veillet, C.; Whipple, A.L.; Wiant, J.R.; Williams, J.G.; Yoder, C.F. (1994) *Science*, 265, 482.

D'Inverno, R. (1992) - *Introducing Einstein's Relativity*. Clarendon Press, Oxford.

Dirac, P.A.M. (1938) - *Proceedings of the Royal Society* **165** A, 199.

Dirac, P.A.M. (1975) - *General Relativity,* Wiley, New York.

Dolgov, A.D. (1983) - in *The Very Early Universe.* Edited by G.W. Gibbons, S.W. Hawking, and S.T.C. Siklos. Cambridge University Press, Cambridge, p.449.

Dolgov, A.D. (1997) - Los Alamos Archives http://arxiv.org/abs/Astro-ph/9708045.

Dolgov, A.D. (1997) *Physical Review D* **55**, 5881.

Eddington, A.S. (1935). *New Pathways in Science.* Cambridge University Press, Cambridge.

Einstein, A. (1923) - in *Principle of Relativity*. Reprinted by Dover, p.111.

Einstein, A. (1923) - ibidem p.177.

Einstein, A. (1921) - *L'Éther et la Théorie de la Relativité*, Gauthiers-Villars, Paris.

Endo, M.; Fukui, T. (1977) - *G.R.G.* **8**, 833.

Eötvös, R.; et al. (1922) *Ann. der Phys.*, 68, 11.

Falco, E. E., Kochanek, C. S., Muñoz, J. A. (1998), *Astrophys. J.* **494**, 47.

Fardon, R.; Nelson A.E.; Weiner, N. (2003) - Los Alamos Archives http://arxiv.org/abs/astro-ph/0309800 v2.

Fermilab, (2000). News Media Contact. Fermilab, 00-12, July 20, 2000.

Feynman, R. P. (1962-3)- *Lectures on Gravitation*, Addison-Wesley, Reading.

Franklin, A. (2000) - *The Road to the Neutrino*, Physics Today, **53**, N. 2, 22.

Fujii, Y. and Nishioka, T.(1991) - *Physics Letters B* **254**, 347.

Fujii, Y.; Maeda, K. -I.; (2003) - *The Scalar Tensor Theory of Gravitation*, CUP, Cambridge.

Frieman, J.A.; et al. (1995) - *Physical Review Lett.* **75**, 2077.

Grischuk, L.P (1975)– Sov.Phys. JETP **40**,409

Grischuk, L.P (1975 b)–Lett. Nuovo Cimento **12**, (2) 60.

Grischuk, L.P (1977)– Ann. N.Y Acad. Sci **302**, 439.

Grøn, Ø. (1986). American Journal of Physics, **54**, 46.

Goldstein, H (1980) - *Classical Mechanics,* 2nd. Edtn., Addison Wesley, N.Y.

Gomide, F.M. (1963) - Nuovo Cimento **30**, 672.

Gomide, F.M. (1965) - *An. Ac. Bras. Ci.* **37**, 425.

Gomide, F.M. (1966) - *Nuovo Cimento* **41**, 156.

Gomide, F.M. (1967) - *An. Ac. Bras. Ci.* **39**, 405.

Gomide, F.M. (1972) - *Nuovo Cimento* **12 B**, 11.

Gomide, F.M. (1973) - *Rev. Bras. Fis.* **3**, 3.

Gomide, F.M. (1976) - *Lett. Nuovo Cimento* **15**, 515.

Gomide, F.M. (1980) - *Lett. Nuovo Cimento* **29**, 399.

Gomide, F.M. (1985) - *Rev. Bras. Fis.* **15**,388.

Gomide, F.M.; Berman, M.S. (1988) - *Introduction to Relativistic Cosmology.* (in Portuguese), McGraw-Hill, São Paulo, 2nd ed. (first edition in 1987).

Gomide, F.M.; Berman, M.S.; Garcia, R.L. (1986) - *Rev. Mexicana Astron. Astrofis.* **12**, 46.

Gomide, F.M.; Uehara, M. (1975) - *Prog. Theoretical Physics* **53**, 1365.

Gomide, F.M.; Uehara, M. (1977) - *Rev. Bras. Fis.* **7**, 429.

Gomide, F.M.; Uehara, M. (1981) - *Astronomy and Astrophysics* **95**, 362.

Guth, A. (1981) - *Physical Review D* **23**, 347.

Halliday, D., Resnick, R., and Walker, J. (2005) *Fundamentals of Physics. 7th.* Edition. Wiley, New York. p. 339. Formula (13-21); etc.

Halverson, N. W. et al. (2002), *Astrophys. J.*, **571**, 604;

Hawking, S.W.(1975) - *Communications of Mathematical Physics* **43**, 199.

Hawking, S.W. (1984) - *Physics Letters B* **134**, 403.

Hawking, S.W. (1993) - *Hawking on the Big-Bang and Black Holes* , World Scientific, Singapore.

Hawking, S. (1996) *The Illustrated A Brief History of Time*, Bantam Books, New York. (1988 earlier non-illustrated edition)

Hawking, S. (2001) - "The Universe in a Nutshell", Bantam, New York.

Horvat, R. (2005) - Los Alamos Archives http://arxiv.org/abs/astro-ph/0505507 v2.

Hoyle, F. (1948). MNRAS, **108**, 372.

Kaplan, D.B. (2004) - Physical Review Letters **93**, 091801.

Katz, J.;Ori, A.(1990) *Class. Quantum Grav.***7**,787.

Kenyon, I.R. (1990) - "*General Relativity*". Oxford U.P., Oxford.

Kolb, E. W.; Turner, M. S. (1990) - *The Early Universe*, Addison Wesley, Redwood City,.

Koyré, A. (1962) - *Du Mode Clos L'Univers Infini*, Presses Universitaires de France.

Kramer, D.; Stephani, H.; MacCallum, M.; Heret, E. (1980). *Exact Solutions of Einstein's field equations*. Cambridge U.P., Cambridge.

La, D; Steinhardt, P.J (1989)–Phys. Rev. Lett **62**, 376.

Landsberg, P.T. (1983) - *Private Communication.*

Lederman, L. (1989) - *Observations of Particle Physics from Two Neutrinos to the Standard Model,* Science, **224**, 664.

Lee, A. T. et al. (2001), *Astrophys. J. Lett.* **561**, L1.

Lemaître, G. (1930). MNRAS, **90**, 668.

Liddle, A.R.; Wands, D. (1992). Physical Review, **D45**, 2665.

Lima, J.A.S.; Carvalho, J.C. (1994) - *GRG* **26**, 909.

Lima, J.A.S.; Maia, J.M.F. (1994) - *Physical Review D* **49**, 5597.

Lima, J.A.S.; Trodden, M. (1996) - *Physical Review D* **53**, 4280.

Lima, J.A.S. (1996) - *Physical Review* **D 54**, 2571.

Linde, A.D. (1988) - *Physics Letters B* **200**, 272.

Linde, A. (1990). *Particle Physics and Inflationary Cosmology*. Harwood Acad. Press, N.Y.

Lynden-Bell, D.;Katz, J. (1985) - *M.N.R.A.S.* **213**,21.

Mach, E. (1912) - *Die Mechanik in Ihrer...* Brockhaus, Leipzig.

Maia, M.D.; Silva, G.S. (1994) - *Physical Review* **D 50**, 7233.

Masiero, A.; Vempati,S.K.; Vives,O. (2004) - *Massive Neutrinos and Flavour Violation*, CERN-PH-TH/2004-142.

McCrea, W.H.(1934-35) - *Zs. für Ap.* **9**, 290.

McCrea, W.H.(1951) - *Proc. Royal Society* **206 A**, 563.

McIntosh, C.B.G. (1973). Physics Letters **A43**, 35.

Misner, C.W.; Thorne, K.S.; Wheeler, J.A. (1973) - see MTW below.

Moffat, J.W. (1995) - *Physics Letters B* **357**, 526.

Morganstern, R.E. (1971). Physical Review, **D4**, 278.

"MTW" - Misner, C.W.; Thorne, K.S.; Wheeler, J.A. (1973) *Gravitation,* Freeman, San Francisco.

Murphy, M.T. et al (2001)– Mon. Not. Royal Astron. Soc, 000, 1-17.

Narlikar, J. (1983). *Introduction to Cosmology*, Bartlett, Boston.

Narlikar, J. (1993). *Introduction to Cosmology*, 2nd. edition (worse than the first one), Cambridge University Press, Cambridge.

Netterfield et al. (2002), *Astrophys. J.* **568**, 38;

Newman, E.T.; et al (1965). Jrn. Math. Phys. **6**, 918.

Novello, M. (1980). *Cosmologia Relativista* in *II Escola de Cosmologia e Gravitação*, ed by M.Novello, CBPF, Rio de Janeiro.

Nordtvedt, K. (1970) Astrophys. J, **161**, 1059.

Ohanian, H.C. (1985) *Physics*, Norton, New York.

O'Hanlon, J.; Tupper, B.O.J. (1972). Nuovo Cimento, **7B,** 305.

Overduin, J.M.; Cooperstock, F.I. (1998) - *Physical Review* **D 58**, 043506.

Özer, M;Taha,M. O.(1986) - *Phys. Lett. B* **171**,363.

Özer, M.;Taha,M. O. (1987) - *Nucl. Phys.B* **287**,776.

Padmanabhan, T. (2002) - Los Alamos Archives http://arxiv.org/abs/hep-th/0212290.

Papapetrou, A.(1974) - *Lectures on General Relativity.* D. Reidel Publishing Company, Boston.

Pathria, R.K. (1972) - Nature **240**, 298.

Peccei, R.D; Solà, J.; and Wetterich, C. (1987) - *Physics Letters* B **195**, 183.

Peebles, P.J.E.; Ratra, B. (1988) - *Astrophysical Journal Lett. Ed.* **325**, L17.

Peratt, A.L. (1990) - The Sciences, N.Y. Academy of Sciences, No. 1, 24.

Perlmutter, S. et al. (1999), *Astrophys. J.*, **517**, 565.

Pimentel, L.O. (1989). Astrophysics and Space Science, **112**, 175.

Physics Today (2004).*Neutrino Oscillation Has Now Been Seen...* , Physics Today, **57**, N. 7, 11, (2004).

Pound, R.V.; Rebka, G.A. (Jr) (1960) - *Physical Review Letters,* **4**, 337.

Purcell, E.M. (1985) - *Electricity and Magnetism*, 2nd. edition, Berkeley Physics Course v2, McGraw-Hill, New York.

Raychaudhuri, A.K. (1975). Physical Review. **53,**1360.

Raychandhuri, A.K (1979) - *Theoretical Cosmology*, Clarandon Press, Oxford.

Raychandhuri, A.K. et al (1992) - *General Relativity, Astrophysics and Cosmology* - Springer, N.Y.

Raychaudhuri, A.K.; Banerji, S.; Banerjee, A. (1992). *General Relativity, Astrophysics, and Cosmology*, Springer-Verlag, New York.

Reasenberg, R.D. (1983) *Phil. Trans. Roy. Soc. (London)*, 310, 227.

Riess, A. G. et al. (1998), *Astron. J.* **116**, 1009;

Rosen, N.(1994)-*Gen. Rel. and Grav.* **26**, 319.

Sabbata, V.de; Sivaram, C. (1994) - *Spin and Torsion in Gravitation*, World Scientific, Singapore.

Sabbata, V.de; Gasperini, M. (1979) - Lettere al Nuovo Cimento **25**, 489.

Salim, J.M.; Waga, I. (1993) - *Classical and Quantum Gravity* **10**, 1767.

Sahni, V.;Starobinski, A. (2000)-*Intl. J.Mod. Phys.* **D9**, 373. Also http://arXiv.org (:astro-ph/9904398 v2).

Sciama, D.N. (1953) - *MNRAS* **113**, 34.

Science, (1992). *Are Neutrinos' Mass Hunters Pursuing a Chimera?*, Science, **256**, 731, (1992).

Science (1992 b). *New Results Yeld no Culprit for Missing Neutrinos*, Science, **256**, 1512, (1992).

Science (1999). *Search for Neutrino Mass...*, Science, **283**, 928, (1999).

Schwarzschild, B. (1992) - Physics Today, **45**, N. 8, 17.

Schwarzschild, B. (1998) - Physics Today, **51**, N. 8.

Schwarzschild, B. (2001). Physics Today, **54** (7), 16.

Schweber, S. (2002) - *Enrico Fermi and Quantum Electrodynamics*, Physics Today, **55**, N. 6, 31.

Sears, F.W.; Salinger, G.L. (1975) - *Thermodynamics, Kinetic Theory, and Statistical Thermodynamics*, Addison-Wesley, New York.

Silveira, V.; Waga, I. (1994) - *Physical Review D* **50**, 4890.

Silveira, V.; Waga, I. (1997) - *Physical Review D* **56**, 4625.

Susskind, L.; Lindesay, J. (2005) - *An Introduction to Black Holes, Information and the String Theory Revolution - The Holographic Universe*, World Scientific, Singapore.

Synge, J.L. (1965) - *Relativity - The Special Theory*, 2nd. edition, North Holland, Amsterdan.

Synge, J.L.; Schild, A. (1978) - *Tensor Calculus*, reprinted, Dover, New York.

Taylor, J.H. ; Weisberg, J.M. (1982) - *Astrophysics Journal,* **253**, 908.

Taylor, J.H. ; Weisberg, J.M. (1989) - *Astrophysics Journal,* **345**, 434.

Tolman, R.C.(1934) - *Relativity, Thermodynamics and Cosmology* , OUP. Reprinted by Dover, New York (1987).

Torres, L.F.B.; Waga, I. (1996) - *MNRAS* **279**, 712.

Uehara, K.; Kim, C.W. (1982) - *Physical Review* **D26**, 2575.

Virbhadra, K.S. (1990) - *Phys. Rev.* **D41**, 1086; **D42**, 2919; **D42**, 1066. See also Aguirregabiria, J.M. et al. (1996) - *Gen. Rel. and Grav.* **28**, 1393.

Waga, I. (1993) - *Ap. J.* **414**, 436.

Wagoner, R.V. (1970)- Phys. Rev. D**1**, 3209.

Webb, J.K., et al. (1999)–Phys. Rev. Lett. **82**, 884.

Webb, J.K; et al.(2001)– Phys Rev. Lett. **87,** 091301.

Weinberg, S. (1972) - *Gravitation and Cosmology*, Wiley, New York.

Weinberg, S. (1989) - *Reviews of Modern Physics,* **61**, 1.

Wrede, R.C.; Spiegel, M. (2002) - *Advanced Calculus*, Second Edition, McGraw-Hill, New York.

Wesson, P.S. (1999) - *Space-Time-Matter (Modern Kaluza-Klein Theory)*, World Scientific, Singapore.

Wesson, P.S. (2006) - *Five Dimensional Physics*, World Scientific, Singapore.

Weyl, H. (1950) - *Space-time-matter*, Dover, N.Y.

Whitrow, G. (1946) - *Nature* **158**,165.

Whitrow, G.; Randall, D. (1951)-*MNRAS* **111**,455.

Whittacker, E. (1935) - *Proc. Royal Society* **149 A**, 384.

Whittacker, J.M. (1966) - *Nature.* **209**, 491.

Will, C.M. (1984) Phys. Rep **113**, 345.

Williams, J.G.; Dicke, R.H.; Bender, P.L.; Alley, C.O.; Carter, W.E.; Currie, D.G.; Eckhardt, D.H., Faller, J.E., Kaula, W.M., Mulholland, J.D., Plotkin, H.H., Poultney, S.K., Shelus, P.J., Silverberg, E.C., Sinclair, W.S., Slade, M.A., Wilkinson, D.T. (1976) *Phys. Rev. Lett.*, 36, 551.

Zel'dovich, Ya. B. (1967) Pi'sma *JETP.***6**, 887.

Zel'dovich, Ya. B. (1967 b) *JETP Lett.***6**, 316.

Zel'dovich, Ya. B.(1968) - *Soviet Physics Usp.* **11**, 381.

Zel'dovich, Ya. B. (1981) *Sov. Phys. Usp.* **24**, 216.

Index

N

O

P

Q

R